Medicine after the Holocaust

Previously published by Sheldon Rubenfeld:
Could It Be My Thyroid?

Medicine after the Holocaust

From the Master Race to the Human Genome and Beyond

Edited by

Sheldon Rubenfeld

In Conjunction with the Holocaust Museum Houston

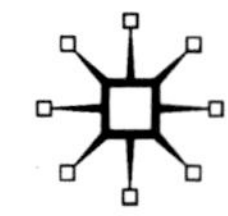

MEDICINE AFTER THE HOLOCAUST

First published in 2010 by PALGRAVE MACMILLAN® in the United States - a division of St. Martin's Press LLC, 175 Fifth Avenue, New York, NY 10010.

Where this book is distributed in the UK, Europe and the rest of the World, this is by Palgrave Macmillan, a division of Macmillan Publishers Limited, registered in England, company number 785998, of Houndmills, Basingstoke, Hampshire RG21 6XS.

Palgrave Macmillan is the global academic imprint of the above companies and has companies and representatives throughout the world.

ISBN: 978–0–230–62192–3 (paperback)
ISBN: 978–0–230–61894–7 (hardcover)

Library of Congress Cataloging-in-Publication Data is available from the Library of Congress.

Design by Integra Software Services

First edition: January 2010

10 9 8 7 6 5 4 3 2 1

Transferred to Digital Printing in 2013

Permissions

Contents

List of Tables and Figures

Tables

Figures

List of Contributors

George J. Annas, J.D., M.P.H., is the Edward R. Utley Professor and Chair, Department of Health Law, Bioethics & Human Rights, Boston University School of Public Health, and professor in the Boston University School of Medicine and School of Law. He is the cofounder of Global Lawyers and Physicians and a member of the Institute of Medicine and the Committee on Human Rights of the National Academies. Professor Annas is the author or editor of 17 books on health law and bioethics.

Michael E. DeBakey, M.D., a pioneer in the field of cardiovascular surgery, was chancellor emeritus of Baylor College of Medicine. Dr. DeBakey also developed the concept of mobile army surgical hospitals (MASH) and was a driving force behind the development of the Veterans Administration Medical Center System and the National Library of Medicine. Dr. DeBakey received numerous medical awards, including the Presidential Medal of Freedom with Distinction, the Albert Lasker Award for Clinical Research, and the Congressional Gold Medal.

Arthur L. Caplan, Ph.D., is the Emmanuel and Robert Hart Professor of Bioethics and chair of the Department of Medical Ethics at the University of Pennsylvania. Dr. Caplan is the author or editor of 25 books, including *When Medicine Went Mad: Bioethics and the Holocaust,* and more than 500 papers in refereed journals of medicine, science, philosophy, bioethics, and health policy.

Jordan J. Cohen, M.D., is president emeritus of the Association of American Medical Colleges and professor of medicine and public health at George Washington University. Dr. Cohen has held many leadership positions in academic medicine, and he is also chairman of the Arnold P. Gold Foundation, which advances humanism in medicine through innovations in medical education.

Francis S. Collins, M.D., Ph.D., is director of the National Institutes of Health and the former director of the National Human Genome Research Institute at the National Institutes of Health, where he led the successful effort to complete the Human Genome Project. In recognition of his contributions to genetic research, Dr. Collins received the Presidential Medal of Freedom, the nation's highest civil award.

Ward Connerly is chief executive officer of Connerly & Associates, Inc., and the founder and president of the American Civil Rights Institute. Mr. Connerly stimulated the University of California to end the use of race as a means for admission in

1995 and has led campaigns in several states to require equal treatment under the law for all residents in public education, public employment, and public contracting.

Theresa M. Duello, Ph.D., is the assistant director of the Diversity Initiatives Endocrinology-Reproductive Physiology Program and associate professor in the Department of Obstetrics and Gynecology at the University of Wisconsin-Madison. She teaches medical school and undergraduate courses on health disparities with an emphasis on the need for medical history to inform the training of future health professionals.

Glen O. Gabbard, M.D., is the Brown Foundation Chair of Psychoanalysis, professor of psychiatry, and director of the Baylor Psychiatry Clinic at Baylor College of Medicine. He has published over 290 scientific papers and book chapters and authored or edited 23 books, including *Psychodynamic Psychiatry in Clinical Practice* and *Psychiatry and the Cinema.*

Henry T. Greely, J.D., is the Deane F. and Kate Edelman Johnson Professor of Law and professor (by courtesy) of genetics at Stanford University. He is also the director of both the law school's Center for Law and the Biosciences and the Stanford Center for Biomedical Ethics Program in Neuroethics. Professor Greely specializes in the implications of new biomedical technologies, especially those related to genetics, neuroscience, and stem cells.

Irving Greenberg, Ph.D., is an ordained Orthodox rabbi and scholar who has been a seminal thinker in confronting the Holocaust as a historical transforming event. A pioneer in Holocaust education, theology, and in Jewish-Christian dialogue, Rabbi Greenberg has published numerous articles and monographs on Jewish thought and religion and was one of the founding figures of the United States Holocaust Memorial Museum.

Michael A. Grodin, M.D., is professor of health law, bioethics, human rights in the Department of Health Law, Bioethics, and Human Rights, Socio-Medical Sciences, Community Medicine, and Psychiatry at Boston University Schools of Public Health and Medicine. He is director of the Project on Medicine and the Holocaust at the Elie Wiesel Center for Judaic Studies at Boston University and is the author of more than 200 scholarly papers and editor or coeditor of five books, including *The Nazi Doctors and the Nuremberg Code: Human Rights in Human Experimentation.*

John M. Haas, Ph.D., S.T.L., is president of the National Catholic Bioethics Center, established in 1972 to apply the teachings of the Catholic Church to ethical issues arising from developments in medicine, the life sciences, and civil law. Dr. Haas was recently named by Pope Benedict XVI to serve as an Ordinary Member of the Pontifical Academy for Life.

Leon R. Kass, M.D., Ph.D., is the Hertog Fellow of the American Enterprise Institute and Harding Professor, Committee on Social Thought at the University of Chicago. He was a founding member of the Hastings Center, has published numerous articles and books on ethical and philosophical issues, and, from 2001 to 2005, served two terms as the first chairman of the President's Council on Bioethics.

Ferid Murad, M.D., Ph.D., is the John S. Dunn Distinguished Chair in Physiology and Medicine, director emeritus of the Brown Foundation Institute of Molecular Medicine, and professor of integrative biology & pharmacology at the University of Texas Health Science Center at Houston. Dr. Murad was awarded the 1998 Nobel Prize in Physiology or Medicine as one of the discoverers of nitric oxide as a signaling molecule in the cardiovascular system.

George Paul Noon, M.D., is professor of surgery in the Michael E. DeBakey Department of Surgery and chief, Division of Transplant and Assist Devices at the Baylor College of Medicine. Dr. Noon has authored numerous book chapters and articles for scholarly and professional journals on subjects that include heart transplantation and mechanical heart-assist devices.

Edmund D. Pellegrino, M.D., is chairman of the President's Council on Bioethics and professor emeritus of medicine and medical ethics at Georgetown University. Dr. Pellegrino, a member of numerous editorial boards, is the author of over 600 published items in medical science, philosophy, and ethics. He is the author or coauthor of 23 books, the founding editor of the *Journal of Medicine and Philosophy*, and recipient of 46 honorary doctorates.

Volker Roelcke, Prof. Dr. med., M.Phil., is the chair and director of the Institute for the History of Medicine at the University of Giessen in Germany. Prof. Dr. med Roelcke is a clinical psychiatrist with a degree in social anthropology. In addition to publishing numerous articles in scholarly journals, he has published four books, including *Twentieth-Century Ethics of Human Subjects Research: Historical Perspectives on Values, Practices, and Regulations.*

Sheldon Rubenfeld, M.D., is clinical professor of general medicine at Baylor College of Medicine. He has taught Jewish Medical Ethics at Baylor College of Medicine for seven years as well as Healing by Killing: Medicine in the Third Reich for three years. Dr. Rubenfeld created the concept for and developed, in conjunction with the Holocaust Museum Houston, Medical Ethics and the Holocaust, which included an exhibit entitled How Healing Becomes Killing: Eugenics, Euthanasia, Extermination and the 17-part Michael E. DeBakey Medical Ethics Lecture Series.

William Seidelman, M.D., is emeritus professor in the Department of Family and Community Medicine at the University of Toronto and an internationally recognized authority on the history of medicine in the Third Reich, especially the role of German and Austrian universities and research institutes. His work has been published in many medical journals including the *British Medical Journal* and the *Journal of the American Medical Association.*

Wesley J. Smith, J.D., a senior fellow in human rights and bioethics at the Discovery Institute, is the associate director of the International Task Force on Euthanasia and Assisted Suicide. He has authored or coauthored 12 books, including *Forced Exit: Euthanasia, Assisted Suicide, and the New Duty to Die*; *Culture of Death:*

The Assault on Medical Ethics in America; and *Consumer's Guide to a Brave New World.*

Avraham Steinberg, M.D., is the director of the Medical Ethics Unit and senior pediatric neurologist at Shaare Zedek Medical Center in Jerusalem and head of the editorial board of the Talmudic Encyclopedia. He has authored 30 books and 240 scientific articles and chapters related to medical ethics. Dr. Steinberg has received numerous awards, among them the Israel Prize in 1999 for his *Encyclopedia of Jewish Medical Ethics.*

Kathryn L. Tucker, J.D., is director of legal affairs of Compassion & Choices, a national non-profit organization dedicated to improving end-of-life care. Ms. Tucker served as lead counsel representing patients and physicians in two federal cases decided by the U.S. Supreme Court in 1997, asserting that competent dying patients have a constitutional right to choose aid in dying. She has successfully defended the Oregon Death with Dignity Act and coauthored the Washington Death with Dignity Act, as well as a number of model legislative measures to improve end-of-life care.

James D. Watson, Ph.D., is chancellor emeritus of Cold Spring Harbor Laboratory. In 1953 Drs. Watson and Francis Crick determined the structure of DNA for which they shared the 1962 Nobel Prize for Physiology or Medicine with Maurice Wilkins. Dr. Watson served as director of the Human Genome Project from 1989 to 1992. He has authored ten books, including *Molecular Biology of the Gene* and the international best seller *The Double Helix.*

Acknowledgments

I am grateful to the Holocaust Museum Houston for accepting my proposal for an exhibit, lecture series, and a book based on the role doctors played in events immediately before, during, and after the Holocaust. In particular, I would like to thank the museum chair, Peter Berkowitz, and executive director, Susan Myers, who chose to devote the museum's resources to this concept. What resulted was a world-class program that eventually touched the lives of thousands of doctors, medical students, and museum visitors over the course of six months. In addition to the thousands who watched the Michael E. DeBakey Medical Ethics Lecture Series (in person or via Internet), more than 50,000 people visited the accompanying museum exhibit, How Healing Becomes Killing: Eugenics, Euthanasia, and Extermination.

Credit is also due to other members of the project's steering committee, including Melissa Brunicardi, Nancy S. Dinerstein, Kelli Cohen Fein, past chair Walter Hecht, Manuel D. Leal, Leo Linbeck III, Cheyenne Martin, Eric Pulaski, Anna Steinberger, Dr. John Thrash, Ileana Trevino, and to current chair Michael Goldberg. The project was aided by the advisory board comprising some of the most influential members of Houston's medical community, and it would never have been possible without the financial support of more than 40 generous donors and supporters.

I would also like to acknowledge the museum staff under the direction of Tamara Savage, who worked tirelessly to orchestrate this enormous project while ensuring the day-to-day operations of the museum continued.

I thank the authors whose works are collected in this volume as well as the other superb speakers who contributed to the program, including Mark Addickes, Edwin Black, Jacquie Brennan, Sandra Carson, Lex Frieden, Michael Gross, Eric Kandel, Susan Lederer, William Monroe, Richard Petty, and Christine Rosen. Many of those who spoke also wrote manuscripts, and my one regret is that there was insufficient space to include all of them in this book. Fortunately, videos of their lectures can be viewed online at http://www.utexas.edu/cola/centers/scjs/med-ethics/Lectures.php. The Schusterman Center for Jewish Studies in the College of Liberal Arts at the University of Texas at Austin established this site to display the videos of all the speakers in the Michael E. DeBakey Medical Ethics Lecture Series.

I thank Arthur Caplan for his singular contributions. He had "been there and done that" with his 1989 conference at the University of Minnesota and publication of collected articles from that conference in *When Medicine Went Mad: Bioethics and the Holocaust.* Art very generously shared his invaluable experience and advice, which enabled me to recruit the highest-quality scholars for the program.

I also thank William Seidelman for his special contributions. Bill was a very early personal supporter and advisor. His many suggestions guided me along an unfamiliar and rocky road and kept me from going into too many ditches. When I didn't understand a historical or political point, I knew I could count on him to explain it.

I gratefully acknowledge the late Michael DeBakey who graciously lent his name to the lecture series and who shared with me his eyewitness accounts of many relevant historical events. The early and consistent support for the project by Richard Wainerdi, Patricia Starck, Glenn Cambor, and Richard Materson is also greatly appreciated.

I am very grateful to my aunt Thea Silber Steuer and to Colonel Josiah C. Wedgwood, a Christian Zionist member of the British parliament, who together rescued my mother eight months after the Nazis marched into Austria. I also thank my cousin René Lehr Steuer, who shared with me the documentation of the rescue.

In addition, I would like to express my appreciation to Wolfgang Fritzsche, my wonderful guide whose historical knowledge and charming ways made my trips to Germany much more rewarding than they otherwise would have been.

A project as large as this one encounters many roadblocks. Ben Tobor, Jeremy Steele, Mark Yudof, Randy Diehl, and Robert Abzug cleared many of them, and I thank them for their timely assistance.

Sheree Bykofsky and Janet Rosen of Sheree Bykofsky Associates, my literary agents, went the extra mile to ensure publication of the book. I gratefully acknowledge their patience, support, and encouragement.

I am grateful to Chris Chappell, my editor at Palgrave, as well as to Samantha Hasey, Kristy Lilas, Sumitha Nithyanandan, and all their staff for providing a steady hand through the editing and publishing processes. I offer a special thank you to Kathy Kobos, my personal and indispensable assistant, for her unflagging energy and tireless devotion to the book and for getting everything just right. I gratefully acknowledge Phyllis Applebaum, Robin Dorfman, and Bobbie Gaspard for reading all of the manuscripts with their sharp red pencils in hand and also Diana Tisdale and Marcella Salas, who offered help whenever help was needed.

Each Wednesday for the past ten years, I have studied Jewish medical ethics with Rabbi Yossi Grossman of the Torah Outreach Resource Center of Houston (TORCH). Rabbis Yaakov Lipsky and Eli Berakah started the class, and Naomi Myers, Alan Winters, Margo Restrepo, Ed Teitel, Stanley Hite, Judith Feigon, Amelia Kornfeld, Muriel Meicler, Jonathan Magid, and many others studied with us. I thank them all for sharing their wisdom with me. I am also grateful to the Baylor medical students in my classes on Jewish medical ethics and on medicine and the Holocaust for sharing their insights with me.

What goes around comes around. More than 20 years ago, Rabbi Irving Greenberg introduced me to Jewish medical ethics at a Wexner Heritage Program seminar. When TORCH arrived in Houston, I asked the rabbis to study Jewish medical ethics with interested physicians including Ron Moses, who not only studied with me nearly every week for the past ten years but also founded an annual Jewish medical ethics conference in memory of his grandfather. When he went to his Wexner Heritage Program seminar, Ron told Rabbi Greenberg about the fruit from the tree the rabbi had planted more than 20 years before. I am grateful for the privilege of studying with both of them.

Finally, I'd like to thank Linda, my loving wife, who fully supported me from the very start of this entire project and who genially managed without my company many nights and weekends. She has been my keenest critic and my best friend, and I am grateful to her for taking on both roles.

Sheldon Rubenfeld
Houston, Texas
April 21, 2009
Yom HaShoah 5769

Foreword

This Past Must Not Be Prologue

Francis S. Collins

In 1907 Eddie Millard was convicted of petty larceny and sentenced to prison in Indiana. There he encountered Dr. Harry C. Sharp, a respected member of the American medical profession. Dr. Sharp proceeded to interrogate Eddie and then informed him that it would be better for society if he were sterilized. Despite Eddie's verbal protests and actual physical struggles in the operating room, a sterilization operation was performed.

Many medical students and physicians today are probably shocked that one of their own colleagues in the honored profession of medicine would carry out this kind of unconsented sterilization on an unwilling prisoner. Yet, in 1907 Indiana had just enacted a law to make such actions legal. That was just the beginning: between 1907 and 1974, approximately 64,000 Americans were forcibly sterilized.

Being associated with the Holocaust and other atrocities of the Nazi regime in Germany, the term "eugenics" carries appropriately dark connotations in the minds of most of us in the medical profession. Less well known are the origins of this pseudoscience in the United Kingdom and the United States. It was Francis Galton, Charles Darwin's cousin, who coined the term "eugenics" in 1883 and made the case for applying selective breeding to the human race. Many scientific and political leaders in America embraced these concepts over the next 20 years. Initially, the focus was on "positive eugenics" programs, such as contests for healthy families and better babies, but slowly the program shifted toward "negative eugenics" in the form of sterilization programs. The mainstream American scientific community played a deeply disturbing role in these events. The Cold Spring Harbor Eugenics Record Office, founded in 1910, carried out surveys of human heredity that went well beyond the boundaries of genetic conditions such as Huntington's disease and claimed to discover Mendelian heritability for all manner of behavioral traits and social maldispositions. Harry Laughlin of the Eugenics Record Office subsequently testified in Congress in favor of sterilization laws and the need to greatly restrict immigration of populations he considered undesirable. The evidence behind his testimony was both scientifically unfounded and overtly racist.

Meanwhile, Germany was reeling from the loss of a substantial proportion of its young men on the battlefields of World War I and was suffering great economic distress as well. Seeking an opportunity to lay blame for Germany's troubles on those less fortunate, Hitler embraced the eugenics mindset, attached it to his own concept of the German master race, and embarked upon a program of state-sponsored eugenics that represents one of the greatest nightmares in human history. Beginning with the involuntary sterilization of some 400,000 Germans, progressing to euthanasia of children with birth defects or mental retardation, moving on to euthanasia of psychiatric patients, and, ultimately, to the massacre of six million Jews, Hitler and his band of genocidal criminals carried eugenics to a depth of evil that still evokes indescribable horror.

Most medical students and physicians know the name of Josef Mengele, the physician at Auschwitz who presided over many horrific medical "experiments," including dissections of living individuals. Yet, relatively few members of the medical profession are aware of the breadth and depth of involvement of German physicians across the country in the Nazi nightmare. By 1942, 40,000 physicians had joined the National Socialist Physicians League and were either directly engaged in the eugenic cleansing program or indirectly involved by their failure to object. Just as we now find it almost impossible to understand how the Christian church could have largely remained silent during this dark era, we must also seek to understand how the mainstream of the German medical profession, supposedly devoted to healing and to the Hippocratic Oath, could have become complicit with such a diabolical program.

It would be a mistake, however, to dismiss this history with self-righteous smugness. Instead, we should ask ourselves whether this could ever happen again. Perhaps there is room for reassurance in most developed countries, at least about state-sponsored eugenics, but there is no room for complacency. After all, widespread abortion of female fetuses has been recently practiced in parts of India and China as a means of gender selection, and presumably medical personnel must be participating.

As for the possibility of eugenic applications to more complex human traits, we can look at the efforts of the Cold Spring Harbor Eugenics Record Office and ridicule the naïveté of those attempts to connect human behaviors and social circumstances to single genes. On the other hand, with the success of the Human Genome Project and subsequent efforts to develop a detailed catalog of human genetic variation, we are increasingly able to identify the heritable components of complex human diseases, and it is only a matter of time before those same approaches begin to identify genetic contributions to human personality, intelligence, and even sexual orientation. While those genetic factors will be individually weak and the environment will play a powerful role in shaping such characteristics, a society that seems willing to embrace a deterministic view of genetics may not always understand those limits.

Perhaps more relevant than state-sponsored eugenics for today's ethical debates, however, is the way in which "homemade eugenics" (a term introduced by Daniel Kevles) is becoming more and more commonplace. Now that preimplantation genetic diagnosis (PGD) is widely available, its application is spreading beyond the efforts to avoid the birth of a child with severe conditions like Tay-Sachs disease and into progressively milder and later-onset conditions. The most dramatic example is in gender selection, which is now offered by 42 percent of in vitro fertilization clinics, simply to provide a couple with the opportunity to choose the sex of their child.

This kind of homemade eugenics provides a classic collision of ethical, legal, and social principles that make it difficult to regulate. After all, most societies in the West generally consider the reproductive choices of couples to be an area that the state should stay out of. On the other hand, society believes that the state has an interest in supporting principles of benevolence, fairness, justice, and non-malevolence. To the extent that homemade eugenics focuses on "the perfect baby" and leads society to be less tolerant of those with disabilities (as may already be happening, for example, for children with Down syndrome), society does have a legitimate interest in overseeing certain reproductive applications of the new genetics.

Fortunately, however, the most frightening scenarios of this kind of homemade eugenics rest upon a mistaken and overly deterministic view of the role of genes. The notion that a couple with unlimited resources could utilize PGD to optimize the traits of a future offspring will quickly run into trouble on scientific grounds. Suppose, for instance, that a couple wanted to optimize five traits for a future child, such as hair/eye/skin color, IQ, athletic ability, body weight, and musical talent. Each of those traits likely has contributions from at least 10 different gene variants. If each embryo had a 50 percent chance of inheriting a particular desirable variant from one of the parents, then getting all 50 of these outcomes "right" would require a thousand trillion embryos from which to select. Even then, the couple would likely be profoundly disappointed with the outcome, since the environment is a huge contributor to nearly all of these traits.

Still, as our understanding of the human genome is increasing at a dizzying pace, it may not be possible to predict all of the potential uses and misuses of this information. Our best defense against a future eugenic nightmare is a sober consideration of the past, coupled with a vigorous and visionary effort to identify future challenges before they turn into crises. That has been the goal of the ethical, legal, and social implications (ELSI) program of the Human Genome Project, which today represents the largest investment in bioethics in human history. That experiment must continue vigorously.

And the medical profession must never forget its role in the eugenic horrors of the past and must be vigilant against the appearance of any evidence of recurrence. This past must not be prologue.

INTRODUCTION

SHELDON RUBENFELD

In 2005 I approached the Holocaust Museum Houston with a proposal for an exhibit and a lecture series about medicine and the Holocaust. The exhibit, How Healing Becomes Killing: Eugenics, Euthanasia, Extermination, was open from September 2007 through February 2008. Concurrently, 33 physicians, biomedical scientists, historians, lawyers, and ethicists discussed the contemporary relevance of the medical profession's behavior in the Third Reich during the Michael DeBakey Medical Ethics Lecture Series. Twenty-one of these distinguished ladies and gentlemen wrote the papers that are included in this book.

Since I am a practicing endocrinologist and neither a bioethicist nor a historian, I feel compelled to explain the events that led to my proposal for the exhibit, the lecture series, and this book. I was unaware of it at the time, but my interest in medicine and the Holocaust began 38 years ago during my internship on "II and IV," the Harvard Medical Service at Boston City Hospital (BCH), one of the most illustrious and highly competitive internal medicine programs in the country at the time. Each year, eight medical students from Harvard and eight from the rest of the country were chosen to man the barricades on II and IV. We were the best and the brightest. Our patients were the poor, addicted, homeless, and unfortunate of Boston, the people other people stepped over while walking on the streets. The interns ran the hospital, calling on upper-level residents only when there was no other choice. The unspoken rule was that you coped with whatever came through the ancient doors of BCH. We were proud, self-sufficient, and omnipotent—truly like gods.

Early in my internship, I admitted a 100-year-old woman from a nursing home with gangrene of her left leg. She had been demented for many years and fed via a nasogastric tube for at least 12 years. She was white-haired, shriveled, dehydrated, and in great pain from her dying left leg.

My fellow interns and I decided that we would put my patient out of her misery with an injection of potassium chloride, painlessly ending her life. The date and time were all set for our act of mercy when our chief medical resident, Richard Cashion, called us together for a conversation.

Who was our chief medical resident? He was, nominally, in charge of us by virtue of his experience and wisdom. But Richard Cashion was from an alien nation, Texas, and, worse, he was religious, with short hair, too. Most of us were from the northeast, wore shoulder-length hair, and believed in the progressive, secular ideas of the sixties. What could he possibly have to say to us?

Dr. Cashion told us that he knew what we were planning. He told us that we would not do it—that we would provide our patient with intravenous fluids, oxygen, pain medications, and antibiotics and that we would harbor no such ideas ever again. I still remember his final words: "Doctors do not kill patients—that is not what we do."

We did not kill our patient, and we never spoke of the incident again. As fate would have it, I subsequently moved to Texas, and years later I asked Dick Cashion if he remembered this particular incident. He did. We talked for a while and then I thanked him for teaching me the most important lesson of my medical career.

Fifteen years after my internship, I was studying "The Third Jewish Commonwealth" with Rabbi Irving Greenberg, one of the contributors to this book, at a Wexner Heritage Program seminar. He made an offhand remark to the effect that a doctor well versed in Judaism would practice medicine differently than one who was not. I did not necessarily believe him, but I was curious and began studying Jewish medical ethics.

Then in 1998, 26 years after my internship, I went to Poland and Israel as an adult participant in the March of the Living, an international educational program that brings Jewish teens from all over the world to Poland on Yom HaShoah, Holocaust Memorial Day, to march from Auschwitz to Birkenau. I stood at the very same spot next to the train tracks at Birkenau where Josef Mengele had stood more than 50 years earlier, directing over 90 percent of the arrivals to the gas chambers and crematoria. I wondered, "What was a doctor doing on the ramps of Auschwitz?"

I visited Germany for the first time in 2004 to find an answer to this troubling question. I went to Hadamar, one of six centers for the "mercy killings" of "lives not worth living" where doctors and scientists developed the techniques of gassing and cremating for the Final Solution. I visited the foremost biomedical research facilities during the Third Reich, the Kaiser Wilhelm Institutes, which were renamed the Max Planck Institutes after World War II. Three institutes played a major role in the medical crimes of Third Reich, including the Kaiser Wilhelm Institute of Anthropology, Human Genetics and Eugenics in Berlin-Dahlem where Josef Mengele sent his research specimens from Auschwitz to his mentor, Dr. Otmar von Verschuer. I also spoke with Gerhard Baader, a child survivor of the Holocaust and a medical historian at the Free University in Berlin, who succinctly summed up Nazi physicians' attitude toward the chronically ill and toward those they considered useless eaters and subhuman: "If you can't cure it, kill it." I immediately thought of my 100-year-old patient.

I couldn't cure her so I was going to kill her. Fortunately, my chief resident gave me a second opinion that stayed my hand. Nazi physicians also received second opinions from America, England, France, Canada, and other Western countries,[1] but those second opinions supported their discriminatory and racist eugenic policies,[2] which ultimately led to the most egregious and well-documented violations of medical ethics ever.

Despite German efforts to cover up their medical crimes, a small library documenting those crimes now exists. German medical and scientific organizations, like the Max Planck Institute, have finally acknowledged their voluntary and enthusiastic participation in Third Reich's policy of "applied biology." On the other hand, America's eugenic past and its support of the eugenic programs of the Third Reich are less well studied and barely acknowledged. American eugenicists, physicians, philanthropists, legislators, and public health officials provided indispensable legislative models, financial aid, and moral support for Germany's racial hygiene programs in the 1920s and 1930s. Hitler rose to power, in part, by co-opting eugenics and combining it with anti-Semitism into a potent political force in the late 1920s. When he became chancellor in 1933, his centralized government accomplished in six years what American eugenicists, restrained by America's federalism, political diversity, and Judeo-Christian heritage, failed to accomplish in six decades—compulsory sterilization, anti-miscegenation legislation, and involuntary euthanasia.

German biomedical science was the most advanced in the world during the first three decades of the twentieth century. The rediscovery of Gregor Mendel's genetic research in 1900 provided a powerful scientific patina and intellectual respectability to the classification of people by race, an idea that quickly became orthodox thinking in the highest circles of academic medicine, government, and philanthropy throughout Germany and the rest of the Western world. In a mere 12 years, the Nazi dictatorship carried racial hygienic thinking to its logical and murderous conclusion, the Holocaust.

America's circumstances today are eerily similar to those of Germany in the early 1930s. American biomedical science is now the most advanced in the world.[3] The Human Genome Project has revitalized a universal interest in biological determinism and eugenics. Patient autonomy and patient rights have become orthodox thinking in the highest circles of academic medicine, government, and philanthropy throughout America and the rest of the Western world. Economic and political power in medicine is concentrated in a strong, centralized American government.

The placement of autonomy, patient rights, and law above other ethical principles, the doctor-patient relationship, and religious moral values is, in part, a consequence of our secular bioethics that originated in the Nuremberg Code. Lisa Eckenwiler and Felicia Cohn, citing bioethicist Jonathan Moreno, describe the danger of the dominance of autonomy over other values this way:

> When it comes to America in the mid-twentieth century, the creation stories are often summarized in a word—"Nuremberg," "Tuskegee," "Willowbrook," and more. All of these stand for part of but not the whole story. In its foundational beliefs in individual autonomy, bioethics had a distinctly American cast. In warning against American "bioethics imperialism," however, Moreno cites the experience of Weimar Germany in the early 1930s when a few physicians created a journal called *Ethics* to discuss various theories of eugenics, then the dominant social philosophy of medicine. Moreno calls the devolution of this journal into a Nazi tract for racial purification, a warning that an intellectual movement can slide into disaster."[4]

While America is qualitatively different from Nazi Germany and while America is not consciously creating another Master Race, our government is, with limited opposition, enabling individuals to practice eugenics through assisted reproductive technologies, abortion on demand, and euthanasia. What opposition there is primarily comes from two quarters: advocates of limited government and those who oppose abortion and euthanasia on religious or ethical grounds.

Although there are many spokespersons for limited government, I feel compelled to briefly discuss the movement toward greater, if not total, control of the United States' health care system by the federal government. If this movement succeeds, then eugenics may be practiced not only by individuals but also by the federal government as its cost for providing health care increases. In particular, the federal government may offer financial incentives to reduce expenditures for terminally ill or severely disabled patients, especially the elderly, by encouraging and financing euthanasia. I find the cautionary words of economist and Nobel Laureate Milton Friedman especially relevant: "The combination of economic and political *power* in the same hands is a sure recipe for tyranny."[5]

The opposition to eugenics in religious as opposed to secular medical ethics does warrant more than a brief discussion. In particular, there are several reasons to examine medical ethics from the traditional Jewish perspective. The first is succinctly stated by Thomas Cahill in *The Gift of the Jews:* "The Jews started it all—and by 'it' I mean so many of the things we care about, the underlying values that make all of us, Jew and Gentile, believer and atheist, tick."[6]

The second reason is the answer to the question *Why the Jews?* posed by Dennis Prager and Joseph Telushkin. They claim that "contrary to modern understandings of anti-Semitism, the age-old Jewish understanding of anti-Semitism does posit a universal explanation for Jew-hatred: Judaism, meaning the Jews' God, laws, peoplehood, and the claim to being chosen."[7] In the case of medicine during the Third Reich, eugenics triumphed over the despised Jewish medical values of the supreme value of all human life and the sanctity of the human body. In America today, autonomy and rights have experienced a similar triumph. A society that believes in the supreme value of all human life will legalize neither abortion on demand nor the right to die. A society that believes in the sanctity of the human body would treat it with respect both during and after life.[8]

According to Jewish tradition, God brought Judaism into the world, in part, to oppose cultures of death or, as it says in Deuteronomy 30:19, "I call heaven and earth today to bear witness against you: I have placed life and death before you, blessing and curse; and you shall choose life, so that you will live, you and your offspring." The original culture of death was Egypt with its Book of the Dead and its monumental pyramids or tombs for its pharaohs and its queens. Hitler's Germany was another culture of death with his *Mein Kampf* and his concentration camps.

Cultures of death are still with us, especially in the Middle East where some political and religious leaders encourage suicide bombing and murder, especially of Jews. Iran, for example, denies the Holocaust, threatens to destroy Israel with nuclear weapons, and supports Hezbollah. Hassan Nasrallah, Hezbollah's secretary general, stated the issue clearly: "We have discovered how to hit the Jews where they are the most vulnerable. The Jews love life, so that is what we shall take away from them. We are going to win, because they love life and we love death."[9]

Statements like these run counter to Islamic bioethics, which agree with Jewish and Christian ethics on the supreme value of human life and the sanctity of the human body.[10] Nonetheless, seven Muslim physicians, all National Health Service employees, attempted simultaneous terrorist bombings in Britain and Scotland in 2007.[11] An Al Qaeda leader in Baghdad commented afterward, "Those who cure you will kill you."[12]

The third reason is that traditional Judaism has, over millennia, created an exemplary system for responding to the ethical challenges of new scientific and technological developments, a system whose foundational language is duty as opposed to rights.[13] This system offers a time-tested model for answering the fundamental questions of who will decide, how will we decide, and which values will prevail.

The fourth and final reason I chose to focus on Jewish medical ethics is that America is a uniquely Judeo-Christian country, the only country created as a democracy with the Bible playing a major role in the process.[14] Judaism was essential to the founding of America, "one of the two wings by which the American eagle flies," the other being the secular philosophy of the Enlightenment.[15] Gabriel Sivan tells the story this way:

> No Christian community in history identified more with the People of the Book than did the early settlers of the Massachusetts Bay Colony, who believed their own lives to be a literal reenactment of the Biblical drama of the Hebrew nation. They themselves were the children of Israel; America was their Promised Land; the Atlantic Ocean their Red Sea; the Kings of England were the Egyptian pharaohs; the American Indians the Canaanites (or the Lost Ten Tribes of Israel); the pact of the Plymouth Rock was God's holy Covenant; and the ordinances by which they lived were the Divine Law. Like the Huguenots and other Protestant victims of Old World oppression, these émigré Puritans dramatized their own situation as the righteous remnant of the Church corrupted by the "Babylonian woe," and saw themselves as instruments of Divine Providence, a people chosen to build their new commonwealth on the Covenant entered into at Mount Sinai.[16]

Because America has a unique Judeo-Christian foundation, because America remains one of the most religious countries in the world,[17] and because the Nazis attempted to destroy Jews and Judaism, it is fitting that the last chapter in this book is about Jewish medical ethics.

Physicians in Nazi Germany betrayed the Hippocratic Oath, the ethical bedrock of the medical profession for more than 2,000 years, when they chose knowledge over wisdom, the state over the individual, a führer over God, and personal gain over professional ethics. If the best physicians of the early twentieth century could abandon their patients, can we, the best physicians of the twenty-first century, be certain that we will not do the same?

This question defies simple answers, but the distinguished authors who contributed to this book address the most pertinent issues in a thoughtful and scholarly manner. With deference to these eminent ladies and gentlemen, I humbly offer my own prescriptions to academic physicians and biomedical scientists so that they, too, can address this vital question.

First, they must study the history of eugenics in Germany and America before the Holocaust. Those who are willing and able should visit the relevant sites in Germany, such as the Hadamar and Bernburg euthanasia killing centers, and concentration camps in Poland, such as Auschwitz and Majdanek. When they return, they should teach what they have learned to medical students and doctors in training. Their studies also should include the origins of the traditional Hippocratic Oath and of bioethics as well as the Judeo-Christian foundation for the moral practice of medicine.

Second, physicians should study and then acknowledge their own profession's eugenic and racist past, just as the American Medical Association recently did when it apologized for its history of racial discrimination against African American physicians.[18] If we are unwilling to speak plainly about our profession's ethical lapses, how can we speak plainly about current medical ethical issues?

Third, physicians should encourage American universities, bioscience laboratories and associations, philanthropic foundations, corporations, governments, and other organizations to make public their involvement in the American and German eugenics movements, as some institutions and government officials have already done. For example, Cold Spring Harbor Laboratory, under the leadership of James Watson, has published in print and on the Web its central role in American eugenics,[19] and the governors of California, Virginia, Oregon, North Carolina, and South Carolina have acknowledged and apologized for their sterilization laws.[20] President Clinton apologized for the Tuskegee Syphilis Study.[21] In 1998 the Massachusetts Institute of Technology (MIT) and Quaker Oats apologized and agreed to pay $1.85 million to children at the Walter E. Fernald State School for feeding them breakfast cereals laced with minute amounts of radioactive iron and calcium tracers in the 1940s and 1950s.[22]

If we follow this three-part prescription, then perhaps physicians will act as courageously as Shifrah and Puah, the midwives whose story is recorded in the Book of Exodus. Shifrah and Puah were instructed by the Egyptian Pharoah to kill the Hebrew baby boys, but they did not do so. We expect the next verse in the Bible to announce the death of Shifrah and Puah for their disobedience, but it says instead, "The king of Egypt summoned the midwives and said to them, 'Why have you done this thing, that you have caused the boys to live!' " Not only are Shifrah and Puah not punished, but the Bible tells us that "God benefited the midwives—and the people increased and became very strong. And it was because the midwives feared God that He made them houses." (Exodus 1:15–21)[23]

NOTES

1. Stefan Kuhl, *The Nazi Connection: Eugenics, American Racism, and German National Socialism* (New York: Oxford University Press, 1994); Mark B. Adams, ed., *The Wellborn Science: Eugenics in Germany, France, Brazil, and Russia* (New York: Oxford University Press, 1990); Angus McLaren, *Our Own Master Race: Eugenics in Canada 1885–1945.* (Toronto: McLelland & Stewart, Inc., 1990); Edwin Black, *War Against the Weak: Eugenics and America's Campaign to Create the Master Race* (New York: Four Walls Eight Windows, 2003).
2. The term "eugenics," which literally means good birth, was coined by Sir Francis Galton to build upon the theory of "natural selection" propounded by his cousin Charles Darwin. Positive eugenics encourages the transmission of more desirable genetic traits by

promoting procreation and medical care for the "superior races." Negative eugenics discourages the transmission of less desirable traits by opposing medical care and promoting birth control for the "inferior races."

3. Part of our success came from replacing the American medical education system of apprenticeships with the German model of medical professors teaching and doing research full time in universities as recommended by Abraham Flexner of the Carnegie Foundation in 1910. This model is, however, the very same one that produced physicians and scientists willing and eager to sterilize hundreds of thousands of their own German citizens and euthanize six million Jews as a public health measure.
4. Lisa A. Eckenwiler and Felicia G. Cohn, *The Ethics of Bioethics: Mapping the Moral Landscape* (Baltimore, MD: The Johns Hopkins University Press, 2007), 5.
5. Milton Friedman and Rose Friedman, *Free to Choose: A Personal Statement* (New York: Harcourt, Inc., 1980), 3.
6. Thomas Cahill, *The Gifts of the Jews: How a Tribe of Desert Nomads Changed the Way Everyone Thinks and Feels* (New York: Doubleday, 1998), 3.
7. Dennis Prager and Joseph Telushkin, *Why the Jews? The Reason for Antisemitism* (New York: Touchstone, 2003), 7.
8. One very public example of disrespect for the human body is physician Günther von Hagens's *Body Worlds* exhibits. Von Hagens flayed corpses, replaced their bodily fluids with brightly colored resins, and posed them in lifelike positions for the purpose of educating adults and "children 8 and older" about the inner workings of the human body. For more discussion on these exhibits, see Michael J. Lewis, "Body and Soul," *Commentary* 123, no. 1 (January 2007), 29–32.
9. Steven Stalinsky, "Dealing in Death," *National Review Online*, May 24, 2004, http://www.nationalreview.com/comment/stalinsky200405240846.asp.
10. Abdallah S. Daar and A. Khitamy, "Bioethics for Clinicians: 21. Islamic Bioethics," *Canadian Medical Association Journal* 164, no. 1 (2001): 60–63.
11. Simon Wessely, "When Doctors Become Terrorists," *New England Journal of Medicine* 357, no. 7 (2007): 635–637. Wessely cites Walter Laqueur, perhaps the foremost scholar of the darkest crimes of the twentieth century, who first observed that doctors were disproportionately represented among the ranks of terrorists. For example, George Habash, the founder of the Popular Front for the Liberation of Palestine and the man behind the aircraft hijackings of Black September, was a doctor. Mohammed al-Hindi received his medical degree in Cairo in 1980, returning to his native Gaza the following year to form Islamic Jihad. Ayman al-Zawahiri, Al Qaeda's number two leader and spokesman, is a surgeon.
12. Deborah Haynes, Michael Evans, and Adam Fresco, "Those Who Cure You will Kill You," *Times Online,* July 4, 2007, http://www.timesonline.co.uk/tol/news/uk/crime/article 2023024.ece.
13. Benjamin Freedman, *Duty and Healing: Foundations of a Jewish Bioethic* (New York: Routledge, 1999).
14. Ken Spiro, *WorldPerfect: The Jewish Impact on Civilization* (Deerfield Beach, FL: Simcha Press, 2002), 245.
15. Michael Novak, *On Two Wings: Humble Faith and Common Sense at the American Founding* (San Francisco: Encounter Books, 2002), 5.
16. Gavriel Sivan, *The Bible and Civilization* (New York: Quadrangle/The New York Times Book Co., 2002), 236.
17. Diane Swanbrow, "U.S. One of the Most Religious Countries," The University Record Online, University of Michigan, http://www.ur.umich.edu/0304/Nov24_03/15.shtml.

18. Robert B. Baker et al., "African American Physicians and Organized Medicine, 1846–1968: Origins of a Racial Divide," *Journal of the American Medical Association* 300, no. 3 (2008): 306–313.
19. Eugenics Archives, Dolan DNA Learning Center, Cold Spring Harbor Laboratory, http://www.eugenicsarchive.org/eugenics/.
20. Dave Reynolds, "Georgia Measure Would Bring Sterilization Apology," *Inclusion Daily Express*, February 5, 2007, http://www.inclusiondaily.com/archives/07/02/05/020507gaeugenic.htm.
21. Sonya Ross, "Clinton's Tuskegee apology also aims to improve relations with blacks,"Associated Press, Tuskegee University, http://www.tuskegee.edu/global/Story.asp?s=1211670.
22. Zareena Hussain, "MIT to Pay Victims $1.85 Million in Fernald Radiation Settlement," *The Tech Online Edition*, January 7, 1998, http://tech.mit.edu/V117/N65/bfernald.65n.html.
23. Nosson Scherman ed., *The Stone Edition Chumash* (Brooklyn: Mesorah Publications, Ltd., 1993), 295.

Part 1

Eugenics, Euthanasia, Extermination

CHAPTER 1

WHEN EVIL WAS GOOD AND GOOD EVIL: REMEMBRANCES OF NUREMBERG

EDMUND D. PELLEGRINO

To do evil that good may come of it can never be the purpose of a man who has not perverted his morality by a false principle; and false principles are not often collected by the judgment as snatched up by the passions.[1]

INTRODUCTION

Samuel Johnson was arguably England's greatest man of letters, but he was no moral philosopher.[2] He was, however, an astute critic of his times and thus, per force, a moralist. Like many moralists he could be mordant, opinionated, and pompous. But he was, as often as not, on the mark in his diagnoses of our human failings. He uttered them in terse and simple language without formal ethical argument. We would surely be mistaken to treat his words as mere banter because of this.

Johnson's words go directly to what I believe went wrong ethically in the practice of medicine in Nazi Germany. It epitomizes the moral pathology that changed physicians from professed healers to professed killers. Johnson's words, in part, answer the questions that still hover over the Holocaust and the other mass murders that accompanied it: "How could it happen?" "How could these physicians justify their betrayal of their commitment to healing?"

These questions continue to vex moral philosophers, psychologists, historians, and most of us. We know these physicians were sane and well educated in medical ethics. Yet, they descended into the hell of moral deprivation quickly and even zealously. We need to know why because individual and mass defection from ethics still occurs among physicians around the world.

My reflections on the crimes prosecuted at Nuremberg are in four parts: (1) a brief review of the psychosocial influences presumed to be behind the physicians' behavior, (2) a summary of their testimony and moral reasoning at Nuremburg, (3) the fragility of medical ethics, and (4) some intimation for the future as professional ethics evolves in the years ahead.

The Psychology of Defection

The question of the psychosocial determinants of physician behavior has been most extensively studied, beginning with the effort by Robert Jay Lifton. All of the proposed explanations are thoughtful, plausible, and possible, but none is entirely satisfactory in light of the enormity of the violation of the physicians' moral responsibilities. An incomplete list of these explanations would include (1) atomization and alienation of private life in totalitarian states,[3] (2) too close an integration of medical interests with National Socialism,[4] (3) psychological "doubling,"[5] (4) a character flaw intrinsic to the German soul,[6] (5) the banality of anti-Semitism in Germany,[7] (6) "situationism" in which bad systems make bad people,[8] (7) desensitization of brain pathways for moral sensitivity,[9] (8) disordered moral reasoning,[10] and (9) some combination of the preceding theories.[11]

Some more recent studies have added further levels of explanation. Ian Kershaw, for example, focused on ordinary Germans.[12] Most of them, he found, were not rabid anti-Semites or "willful executioners." The "executioners" in James Waller's opinion were the police.[13] Although most Germans were aware of the fate of their Jewish neighbors, Kershaw judged them to be indifferent to that fate.

Waller concentrated on ordinary people, but his analysis was framed in terms of evolutionary biology. He invoked a biological tendency of man and some higher primates to engage in extreme and systematic violence against their fellows. As evidence, he cites Holocaust-like activity in Cambodia, East Timor, Kosovo, Rwanda, and Darfur. From American history, he selects the mass killing of Native Americans at Sand Creek, Colorado. Waller doubts we can do much about this biological tendency to depersonalize those different from us. More specifically, he attributes genocide to this depersonalization of the "other." The biological facts, he suggests, are reinforced by the social construction of our world as well as by where we are and who we are.

The sheer diversity of these psychosocial explanations calls for some empirical attempt at a unified theory. To this end, relating psychopathology to defective moral reasoning is a necessary first step.

The "Rationality" of Defection

Up to this point, for heuristic purposes, I have intentionally separated psychosocial factors from their rationalized explanations. Obviously, emotions, feelings, and values cannot so neatly be disentangled from one another or from reason. Indeed, Johnson's words suggest they are linked when he says that false principles are "snatched up" by the passions.

Several years ago, David C. Thomasma and I wanted to know how such highly educated physicians could reconcile their participation in the experiments laid bare at Nuremberg.[14] We looked for an answer in the defendants' own words when they were examined on the witness stand.

All the defendants appeared to be mentally competent based on their understanding of and responses to questions asked of them. They were "rational" in the sense of comprehending the nature of the questions and giving relevant answers. They seemed to realize that the charges were both legal and ethical. In general, their primary defense was the argument that certain ends and certain procedures as a means to those ends were legal and, therefore, ethical. Thus, they argued, physicians were required to comply. By complying, the physicians thought they were relieved of all moral guilt.

Following legal or military orders, for the German doctors, superseded any personal judgment that an order might be immoral or mistaken. Clearly, the interests of the state took precedence over those of individual citizens. Moreover, for the German doctors, the state's sovereignty overrode traditional medical ethics as well as eternal and natural law. If an experiment ended fatally, the responsibility fell on the state, not the doctor. Finally, evil was turned to good in the minds of the experimenters. One defendant even said he had acted out of "pity for the patients."

Ulf Schmidt has published a detailed study of Karl Brandt, the leader among the defendants and Hitler's personal physician. Brandt led the T-4 program of involuntary euthanasia for the "unfit" as well as the program of experimentation on Jews and political prisoners. Schmidt found no evidence of evil in Brandt's early life. Brandt was, however, heavily influenced by Hitler's *Mein Kampf* and worldview. Brandt protested repeatedly that "pity" for the patient was his chief concern. Schmidt commented that a history like Brandt's cannot be ignored because "we cannot survive its repetition."[15]

These studies tell us little about which of the psychosocial or ethical factors weighed most heavily, individually, or generally on the perfidy of the Nuremberg defendants. There is, thus, at present, no completely satisfactory answer to the question, "How could it happen?" What is clear is that in their personal defense statements, the indicted physicians made betrayal of the physicians' ethics justifiable for the good of *Volk* and *Reich* (people and empire). They placed the good of National Socialism and national survival ahead of their moral obligations to the sick, the innocent, and the vulnerable.

That ideology was embedded in a theory of racial purity whose aim was a "final solution" by way of genocide—first of the Jews. Following that, as one of the exhibits in the Holocaust Museum Houston showed, there were plans to extend racial purification to other groups, presumably, until the world would be rid of Slavic peoples, homosexuals, and others. The culmination of a biologically based ideology of "unfit people" was the German death camp. The ideology dictated that such people should be eliminated for their own good and the good of the world. If they were not bred out of existence, they had to be otherwise eliminated.[16]

THE FRAGILITY OF THE HIPPOCRATIC OATH

The first reaction of many people, and especially of physicians, on hearing the details of the Nuremberg trials was, "How could this happen?" Didn't the German physicians take the Hippocratic Oath? Would this not have demanded, at the least, nonparticipation or protest by the German medical profession? Should not all the world's physicians have raised their voices against such diabolical acts?

This is a question for another day. Suffice it to say that some German physicians did protest. But as Naomi Baumslag demonstrates in her book *Murderous Medicine*,[17] the Nazis mounted an extensive program of deception to hide the purposes of the death camps. This neither exonerates nor explains away the weakness of the voices of protest worldwide. But it was difficult to mount a protest when the facts were so falsified, obscured, and destroyed. The difficulties do not constitute an excuse, however.

So far as the German physicians go, the Hippocratic Oath seemed not to be a deterrent of any significance for those who participated. Yet, for centuries, the first moral principle of the oath had been the principle of beneficence, especially forbidding the taking of life. The oath was overridden by what the Nazi physicians considered a higher morality, the law of the Third Reich, the economic and physical survival of the German *Volk*, and the exigencies of war

The German physicians knew the oath, and they were taught medical history more extensively than physicians in most other countries. Indeed, one of the defendants was author of a popular book on medical ethics. The rejection or distortion of the oath reveals that in a world in which naked power prevails, it counts for little. This was true in Soviet Russia as well. The only strength of the oath is moral. To be effective, the oath requires a community of physicians who support its precepts whatever hardship that might impose. An oath is a solemn promise—not a mission statement, a charter, or a set of ideals simply to be emulated in the interests of group identity.

If the oath is to be fully convincing, it must be grounded on a firm pediment of solid ethical argumentation. Long life and wide acceptance of the oath will not suffice to give it moral force. The fragility of the oath imposes on the profession the task of providing a firmer grounding in ethical principles and an act of undeviating personal commitment to the health and welfare of the sick.[18] Sadly, the reverse trajectory is the reality. The Hippocratic Oath and ethic are being eroded precept by precept. For example, the principle of non-maleficence has replaced the principle of beneficence; the prohibitions against abortion and euthanasia are no longer recognized. Sexual relationships with patients are, for many, simply a matter of "consenting adults" ignoring the intricacies of the physician-patient relationship. Confidentiality has been weakened. In this age of autonomy and individual choice, some physicians simply assert that one may refuse to take the oath.

I do not wish to argue for one or the other of these precepts. The point here is that the Hippocratic corpus has been made even more fragile than it was when devastated by the ethics of National Socialism.

Some Intimations for the Future

What lessons can we learn for medical ethics today from the monstrous perfidy of the Nuremberg physicians? Even to respond to such a question presents an ethical challenge. Some believe that to respond would in some way dishonor the victims because taking the arguments of the defendants seriously might legitimate them in some way. In addition, there are the dangers of all arguments by analogy. One such danger is depreciation of the uniqueness of the Holocaust along with the atrocities revealed at Nuremberg. Another is the assumption that history would repeat itself in the same way. Finally, literal comparisons with a variety of other massive depredations of human dignity might be made. The Holocaust and the Nuremburg trials are too often used as a rhetorical device to deprecate ideas or persons who may be unworthy for a variety of reasons not related to Nuremburg or the Holocaust.

Granting these limitations, the fact is that the physicians tried and convicted at Nuremberg were rational human beings, in positions of power. Had they remained faithful to their solemn oath as healers, most of the medical atrocities could not have taken place. Moreover, those physicians were moral agents: they were sane, capable of making choices on moral matters, and capable of being responsible and accountable for the results of their choices. They were, in short, capable of refusing—singly or en masse—as a matter of conscience. That they did not do so resulted from a complex of reasons that this reflection has examined only in small part.

One thing is certain from the Nuremberg testimony: the defendants were convinced that they had acted for the "good." The evil consequences were necessary if the "good" was to be achieved. They acted out with precision the moral charade that Johnson defined so pithily. Instead of being advocates for the fellow human beings presenting themselves before them, the Nuremberg physicians became their torturers and murderers.

To be sure, the Hippocratic Oath proved to be a fragile instrument. It should not be eliminated or revised as so many have done and are doing today to make it more amenable to our contemporary ethos. The revisions, needless to say, add nothing of moral depth that might have prevented the experiments or the death camps. They offer pragmatic accommodation in place of firm moral norms. I have written elsewhere about what needs to be done to fashion a moral philosophy to undergird the physicians' oath—one that is based in the formation of character in the physician and one that grounds the "physician's" ethics in the unchanging realities of the relationship between one human being, vulnerable and in distress, and another human being dedicated to helping and healing.[19]

We must also become alert to the earliest intimations of danger to patients even in democratic societies by the erosion of the morality of the physician-patient relationship—for example, making the physician a guardian of society's resources rather than his patient's welfare, denying the right of conscientious objections on the part of the physician, and confusing what is legal with what is ethical. The morally "neutral" physician is a scandal to morality and humanity. Public policy must look first to its impact on the welfare of those it promises to serve. All humans are entitled to respect, and none is to be disvalued. Toleration of "collateral moral damage" to the innocent to attain some putative good is perverting morality to "false" principles. This is just what Johnson and many other moralists have warned us against.

Remembrance of Nuremberg is a moral obligation lest we forget how easily evil becomes good and good becomes evil.

The recollection of the past is only useful by way of provision for the future.[20]

NOTES

1. Samuel Johnson, *The Quotable Johnson, A Topical Compilation of his Wit and Wisdom*, comp. Stephen C. Danckert (San Francisco: Ignatius Press, 1992), 87.
2. Charles H. Hinnant, *Samuel Johnson, An Analysis* (New York: Palgrave Macmillan, 1988).
3. Hannah Arendt, *The Origins of Totalitarianism* (New York: Meridian Books, 1971).
4. Robert N. Proctor, "Nazi Doctors, Racial Medicine, and Human Experimentation," in *The Nazi Doctors and the Nuremburg Code*, ed. George J. Annas and Michael A. Grodin (New York: Oxford University Press, 1992), 17–31.
5. Robert J. Lifton, *The Nazi Doctors: Medical Killing and the Psychology of Genocide* (New York: Basic Books, 1986), 418–429.
6. Daniel J. Goldhagen, *Hitler's Willing Executioners: Ordinary Germans and the Holocaust* (New York: Vintage, 1997).
7. Michael Burleigh, *Ethics and Extermination: Reflection on Nazi Genocide* (New York: Cambridge University Press, 1997).
8. Stanley Milgram, *Obedience to Authority: An Experimental View* (New York: Harper Perennial Modern Classics, 2004); see also Philip G. Zimbardo, *The Lucifer Effect: How Good People Turn Evil* (London: Rider & Co., 2008).
9. See Walter Glannon, "Moral Responsibility and the Psychopath," *Neuroethics* 1, no. 3 (October 2008): 158–166.
10. E. D. Pellegrino and D.C. Thomasma, "Dubious Premises—Evil Conclusions: Moral Reasoning at the Nuremberg Trials," *Cambridge Quarterly* 9, no. 2 (2000): 261–274.
11. Arthur L. Caplan, "The Doctors' Trial and Analogies to the Holocaust in Contemporary Bioethical Debates," in *The Nazi Doctors and the Nuremburg Code*, ed. George J. Annas and Michael A. Grodin (New York: Oxford University Press, 1992), 258–275.
12. Ian Kershaw, *The Germans and the Final Solution* (New Haven: Yale University Press, 2007).
13. James E. Waller, *How Ordinary People Commit Genocide and Mass Killing*, 2nd ed. (New York: Oxford University Press, 2007).
14. Pellegrino and Thomasma, "Dubious Premises—Evil Conclusions," 261–274.
15. Ulf Schmidt, *Karl Brandt: The Nazi Doctor: Medicine and Power in the Third Reich* (London: Continuum, 2008), 400.
16. Elof Axel Carlson, *The Unfit: A History of a Bad Idea* (Cold Spring Harbor, NY: Cold Spring Laboratory Press, 2001).
17. Naomi Baumslag, *Murderous Medicine: Nazi Doctors, Human Experimentation, and Typhus* (Westport, CT: Praeger, 2005).
18. Edmund D. Pellegrino, "Professional Codes," in *Methods in Medical Ethics*, ed. Daniel P. Sulmasy and Jeremy Sugarman (Washington, D.C.: Georgetown University Press, 2001), 70–86.
19. Edmund D. Pellegrino, "Professing Medicine, Virtue Based Ethics, and the Retrieval of Professionalism," in *Working Virtue: Virtue Ethics and Contemporary Moral Problems*, ed. Rebecca L. Walker and Philip J. Ivanhoe (New York: Oxford University Press, 2007), 61–85.
20. Johnson, *The Quotable Johnson*, 83.

CHAPTER 2

MEDICINE DURING THE NAZI PERIOD: HISTORICAL FACTS AND SOME IMPLICATIONS FOR TEACHING MEDICAL ETHICS AND PROFESSIONALISM

VOLKER ROELCKE

Since the early 1980s, much historical research has been done on medicine during the Nazi period.[1] In general, this research has focused on three issues:

1. The relationship between physicians and the state, with two subheadings: the relationship between German physicians and the new regime and the fate of Jewish and socialist physicians under Nazi policies.[2]
2. The eugenic and racial health policy, which resulted, among other things, in the compulsory sterilization of at least 360,000 individuals and the killing of almost 300,000 psychiatric and mentally handicapped patients.[3]
3. The atrocious human subjects research in concentration camps and elsewhere.[4]

In the last few years, a fourth issue has come into view: forced labor in medical institutions and the fate of diseased forced laborers.[5]

Before describing some of the results of this historical research on the Nazi period itself and discussing their implications for medicine today, a few sentences are needed on the post-World War II handling of the Nazi past in German medical schools. Although, today, in-depth accounts exist for about one-third of the medical schools in the Nazi period, most of them have only been published within the last ten years.[6] Only one of these publications dates from the first half of the 1980s:[7] a history of

Giessen University Medical School. It is highly significant that this early volume was initiated and edited neither by representatives of the medical school itself, nor by professional medical historians, but by a small group of students of medicine and political science. The publication was not at all welcomed by the medical school and the university, and it is perhaps no accident that the four authors eventually pursued their careers outside academic institutions.

Almost all other publications on medical schools date from the late 1990s or after the year 2000—comparatively late in the context of the broader historiography of Nazi medicine. This fact points to a highly ambivalent attitude of medical school representatives toward researching their own past.

The hesitancy of academic physicians to reflect on their past—that is, to reflect on medical atrocities committed within their own institution—is particularly in need of explanation when confronted with the readiness of the vast majority of German physicians to comply with the expectations of the Nazi regime and to cooperate with it even at the cost of endangering basic human rights.

New Historical Approaches

The four areas of research mentioned at the very beginning—physicians and the state, eugenics and racial hygiene, human subjects research, and forced labor—were not separate issues, but were interrelated in manifold and complex ways. For example, new career opportunities were created by the expulsion of a considerable proportion of professionals from their posts and by newly emerging institutions and fields of research. These opportunities were, in turn, related to the nationwide implementation of the program for enforced sterilization to prevent the spread of hereditary disorders, a program aimed at improving the health of the national organism *(Volkskörper)*, which, in turn, was a notion used synonymously with that of the German race.

These new career opportunities may have been one motivation for the high rate of support of the new regime by the majority of physicians. These career opportunities cannot clearly be separated from the changing contents of research, in particular from the rapidly growing focus on issues of heredity in all medical specialties.[8]

For example, in the field of psychiatry, the forced migration of Jewish and socialist physicians was closely linked to the marginalization or exclusion of a number of etiological concepts. In particular, the neglect of sociological or psychodynamic explanations for the causation and course of disorders went hand in hand with the rapid stabilization and hegemony of somatic and, in particular, eugenic-genetic interpretations of psychiatric conditions. The massive financial and symbolic investment in the eugenic-genetic paradigm that dominated the programs, practices, and funding policies of psychiatric research from 1933 onward caused young physicians and scientists to concentrate their activities on exactly these issues. Consequently, in this field of scientific enquiry, empirical data were collected, hypotheses were generated, and an increasing demand for further research was created, while competing approaches to understanding and managing psychiatric disorders became more and more marginal.[9]

Thus, I want to argue that the aims and practices of medical professionals during the Nazi period can only be understood adequately in the context of (1) the specific challenges and opportunities created by the political system and (2) the explicit and implicit values attached to health and physical performance in the service of the nation or race.[10]

From its coming to power in 1933 onward, the regime aimed to promote those supposedly "fit" to contribute to the highly valued *Volkskörper*, while simultaneously segregating and, in the end, eliminating those who appeared to endanger the efficiency of national economic efforts or the health of the race, understood as a collective organism. It was this rationale that dominated the health and science policies of the Nazi regime in the years following its coming to power.[11] Therefore, nationwide preventive screening programs to fight hereditary diseases were implemented, and research to scientifically justify these programs was funded. After this first reconfiguration in 1933, the onset of World War II in 1939 again brought new challenges and modified ramifications. The new challenges were increasing economic pressures, a perceived need to divert resources to the various branches of the army, and the necessity of simultaneously caring for both wounded soldiers and the civilian population, resulting in new priorities for the distribution of resources in health care and medical research.[12] The systematic clearance of mental asylums through the program of patient killings, or euthanasia, followed these perceived needs and rationality. The euthanasia program allowed for both the radical purification of the genetic composition of the national body and the possibility of eliminating the economic burden supposedly caused by chronic psychiatric and handicapped patients. And finally, it created plenty of hospital space for wounded soldiers and civilians affected by the war.

The needs of the army and air force also justified, in the perspective of many physicians, extreme and atrocious forms of human subjects research on prisoners in concentrations camps, in mental asylums, and in hospitals in the occupied territories. For example, high-altitude experiments in the context of aviation medicine were conducted on prisoners in Dachau, and experiments to identify the organism causing hepatitis were conducted on prisoners in Sachsenhausen and elsewhere.[13]

MYTHS AND REALITIES: RESULTS OF HISTORICAL RESEARCH

By adopting such an analytical approach, recent historical research has put into question a few myths regarding the relation between the Nazi regime and medicine. These myths may be summarized as follows:

1. Medical atrocities were committed by a few fervent Nazi doctors and were the result of irrational ideologies imposed from above by Nazi politicians on apolitical physicians.
2. The programs of mass sterilizations and patient killings were the expression of an ideology that had little or nothing to do with contemporary state of the art medical reasoning and practice.
3. The activities of eugenically motivated medical geneticists and medical scientists experimenting on humans in concentration camps had nothing to do with

science, but were rather the expression of racial ideology or even individual perversion disguised as science and, therefore, better termed "pseudoscience."

I intend to deconstruct these myths by using the results of recent medical historical research. The intent of deconstructing the myths is not to construct a revised, more positive picture of medicine during the Nazi period. Rather, the intent is to demonstrate significant similarities in basic conceptions and behavior patterns between medicine in the Nazi context and medicine before and after it. To put it differently, the atrocities of Nazi medicine were not something that was restricted to Germany and the period between 1933 and 1945; rather, they were the extreme manifestation of some problematic potentials and tendencies implicit in modern medicine in general, which appeared gradually in the context of a totalitarian state and accelerated rapidly during World War II.

The First Myth

It has now been clearly established that a majority of German physicians and biomedical scientists, even if they were not party members, were prepared to accept the promises and temptations of the new authorities. This is remarkable for a professional community whose members, at the time, were perceived as international leaders within their respective medical disciplines.[14] Physicians and medical scientists accepted these promises and temptations for a number of reasons. Young physicians could improve their chances of getting jobs; physicians already in junior positions could advance to more secure careers; and the senior staff in tenured positions could stabilize and increase their influence in their respective fields, receive additional research funds, or increase the impact of their scientific expertise on social and health policies.

The percentage of physicians who became members of the Nazi party or related organizations, such as the SS, is very indicative of this willing behavior. Depending on the specific region of Germany, approximately 50–65 percent were members.[15] The implication is twofold. First, between 35 and 50 percent of physicians did not join the party—a clear indication that nobody was forced to do so. While there is ample documentation that party membership made it easier to start or continue a career, many physicians were appointed to permanent positions, including university chairs, without being a member of the party, as long as they abstained from public critique of Nazi policies or politicians. A case in point is Ferdinand Wagenseil, who was appointed full professor of anatomy in Giessen in 1937 in spite of the fact that he was not a party member.[16]

On the other hand, 50–65 percent is a much higher percentage of membership than in other academic professions, such as law or teaching, which indicates that physicians had a specific affinity for the Nazi regime. This affinity is most likely due to at least two factors. First, there was an increase in jobs and career opportunities after 1933 because of the expulsion of Jewish and socialist physicians, the creation of a nationwide system of hereditary health courts, and the related research and educational activities. Second, the promises and temptations of the regime's program of "applied biology" (i.e., the framing of its social and population policies in biological and medical terms) meant that physicians became one of the most cherished professional groups of the regime.[17]

The perceived or real force exerted by the regime is an insufficient explanation for the actual behavior of physicians. Not only regarding party membership but, on almost all levels of medical activities, there existed some (occasionally considerable) range of possibilities for action. As long as physicians did not belong to the stigmatized and later persecuted minorities, in particular Jews or "socialists," there was a range of possibilities of action in addition to rejection of party membership. This range of possibilities is exemplified by the actual behavior of different groups of physicians who, according to the Law for the Prevention of Hereditary Ill Progeny (law of forced sterilization), were all obliged to notify the newly established hereditary health courts about every individual with suspected hereditary illness. Whereas psychiatrists in public service were very compliant in carrying out this legal demand (they depended for their job and career opportunities on state institutions and their representatives), psychiatrists in private practice were not at all compliant. Thus, a recent instructive study has documented that in a Bavarian district around Erlangen-Nuremberg where membership in the Nazi party was particularly high (approximately 75 percent), physicians and psychiatrists in private practice made no notifications according to the sterilization law for the whole period between 1934 and the beginning of the war in 1939. The explanation for this behavior is certainly not that all these physicians were political dissidents or even involved in active resistance against the regime. Rather, this phenomenon is an expression of the fact that these physicians relied on a good and trusting doctor-patient relationship for their income, and they did not want to jeopardize that relationship.[18] These findings illustrate that the demands on physicians by the sterilization law could be ignored without negative consequences.

The possibility of refusing to meet the expectations of the regime was also present in the euthanasia program. For example, Gottfried Ewald, a professor of psychiatry in Göttingen, declined to act as an expert reviewer in the formalized decision-making process to choose patients for "euthanasia." Ewald not only declined but wrote letters to various functionaries arguing against the whole program. Nothing negative happened to him: he was neither put in jail, nor was he dismissed from his position.[19]

These and other examples document that there was a considerable range of possibilities for noncompliance. It is a sad fact that the vast majority of physicians did not take advantage of, or even explore, this range of possibilities. Rather, the majority of physicians chose to comply.

THE SECOND MYTH

The notion of a "range of possibilities for actions" still presupposes indirectly, however, that all inhuman activities in medicine were forced on physicians "from above" and that physicians had somehow to comply. Although this is not completely wrong, it is only one dimension of the story and leads to the second myth.

Recent historical research has documented that the initiative for almost all intrusive or atrocious medical activities came from physicians and biomedical scientists themselves. This was true for the programs and organizational blueprints for the eugenically motivated forced sterilization; for the mass killing of psychiatric and mentally handicapped patients considered "unworthy of living" in the context of the euthanasia program; and for almost all the known cases of human subjects research in psychiatric and other asylums, concentration camps, or hospitals of the German-occupied territories. It was also true for the "use" of forced laborers in medical institutions where they had to carry out construction, kitchen work, and, probably, auxiliary laboratory work. The only physicians who were forced to participate in problematic or atrocious activities were those who were themselves prisoners (e.g., in concentration camps).

The concepts of eugenics and racial hygiene do not have their origins in the Nazi period, nor were they invented by Nazi politicians and forced upon physicians. Both concepts and terms have their origins in the late nineteenth century, and they were coined and popularized by physicians, biologists, and statisticians.[20] By the end of the 1920s, eugenic or racial hygienist ideas motivated most of the research in medical genetics, which was funded by the largest scientific and philanthropic agencies, such as the Rockefeller Foundation. For example, the Rockefeller Foundation provided most of the money for a nationwide anthropological genetic survey of the German population in the years from 1929 to 1934.[21] Prominent medical scientists, such as Eugen Fischer and Ernst Rüdin, developed eugenic and racial hygienic ideas in Germany and then, only a few years later, closely cooperated with the Nazi regime to implement its sterilization policies. Although Rüdin's research in psychiatric genetics after 1933 was aimed at establishing a scientific foundation for Nazi racial hygiene policies, he was still considered a leading figure in international scientific circles and was, for example, invited to be a plenary speaker at the World Congress of Genetics in Edinburgh as late as 1939, immediately before the onset of World War II.[22]

Similarly, the idea that it might be possible through medical expertise to distinguish between human life worthy of living and life unworthy was not invented by Nazi politicians but by renowned physicians and lawyers many decades before the Nazis came to power.[23] In times of economic crisis, like those immediately after World War I and after the world financial crisis in 1929, such ideas met with particular resonance among physicians and the wider public. In 1939 psychiatrists and pediatricians took the initiative to draw out and implement a program to systematically kill psychiatric patients and handicapped children, a program that was implemented with the cooperation of Nazi governmental institutions.

Regarding medical research, a case in point is Hans Schaefer, one of the most prominent physiologists in West Germany in the 1950s and 1960s. Schaefer was firmly rooted in Catholicism and had considerable reservations about the Nazi regime's policies. Nevertheless, from 1940 onward, he framed his research on cardiorespiratory and ophthalmic physiology to match the expectations of the German Air Force. His research, if successful, would lead to additional funding and exempt himself and members of his research team from immediate military service at the front. Schaefer repeatedly approached the relevant political and military institutions to offer his services and those of his institute in Bad Nauheim (the Kerckhoff Institute associated with Giessen University). Although his approaches were initially turned down, he finally obtained money for research projects related to the physiology of the heart and eye under the extreme conditions to which air force pilots were exposed[24] and, thereby, joined the network of physiologists who supported the German Air Force in times of war.[25]

At a more general level, this kind of behavior may be described and understood by using the concept of "career resources."[26] The Nazi state by no means terminated career opportunities in medicine and the biomedical sciences. Instead, it attempted to completely reconfigure career pathways and possibilities to attract young and promising physicians and scientists into those fields of medicine and biomedical research that were of highest priority to the regime. Areas of interest included hereditary and racial health, microbiology and hygiene, as well as physiology to enhance the performance of human laborers, soldiers, and officers in war and war-related industries.[27]

The war, furthermore, created a specific configuration of motives. Even when one disagreed with the political aims of the regime, cooperation with the military was seen as justified for at least two reasons: (1) It was difficult to refuse service to the nation and the community in times of external danger, particularly when young men were risking their lives at the front. Who could stand aside and deny these young men medical, scientific, and technological expertise? (2) War-related research would save oneself and, potentially, one's research colleagues from military service at the front.

THE THIRD MYTH

While some of the areas pursued by Nazi human subjects researchers were either outdated (e.g., military surgery[28]) or controversial (e.g., microbiology), many were in tune with, or even at the forefront of, the research pursued by the international scientific community (e.g., aviation medicine and medical genetics[29]). The methods and techniques employed also represented a broad spectrum, from the traditional, even obsolete, to the innovative in medical genetics, biochemistry, psychiatry, and especially physiology. Therefore, these research activities cannot simply be classified as "pseudoscience" or mere irrational activities carried out under the guise of medicine. In most cases, however, the practical implementation of the research methods and techniques in question was brutal with a total disregard for the suffering of the research subjects, who frequently were inmates of either concentration camps or asylums for psychiatric or mentally handicapped patients. The vast majority of the human subjects researchers followed the intrinsic logic of their scientific disciplines and, at the same time, enjoyed unrestricted access to human subjects as a consequence of the Nazi regime's policies during wartime. Thus, they were able to do research that otherwise could not have been done, such as the correlation of data from living human subjects with data from the same subjects immediately after they had been killed.

Both before 1945 and at the Nuremberg Doctors' Trial, some of the researchers formulated their own explicit arguments to justify or defend their activities. This kind of moral justification was not completely absurd and merits closer attention.[30] One of the central arguments was the high value of gaining new scientific knowledge to enhance, for example, labor performance, and to help future generations of patients. Another argument was the priority of the well-being of society as a whole, or the "body national" *(Volkskörper)*, over the well-being of the individual, especially in national crises. A critical analysis of these justifications and the impact of similar arguments in today's debates (e.g., what kind of biomedical research is justified in the context of the "war against terrorism") is a necessary task for both historians of medicine and bioethicists.[31]

Conclusions for Today: Medical Ethics and Education

The attitudes, motivations, and actual behavior of physicians and biomedical scientists during the Nazi period as reconstructed in recent historiography suggest a few points for broader deliberations on medical ethics and professionalism. Most of these attitudes and behaviors appear not to be Nazi specific, but rather the expression of underlying widespread attitudes of medical professionals and biomedical researchers.

First, the difficulties in resisting the promises and temptations of those in power are ubiquitous. The sources of power may be located within the medical schools or within other institutions, such as the army, the state, or organizations that distribute funds. Much of the behavior of physicians in the Nazi period may best be understood as a result of failure to resist these temptations, rather than as a consequence of immediate pressure from the regime.

Second, the value hierarchies identified in Nazi biomedical sciences (priority of the pursuit of scientific interests above all other considerations, with as few restrictions as possible) are not Nazi specific, either. The Nazi regime was, in this respect, the ideal context for researchers (one might provocatively speak of a context of "deregulated" research) since it provided ample human subjects for research unconstrained by any legal and moral limits.

Third, scientific and health policy programs that aim at the enhancement of human performance and the improvement of the genetic constitutions of individuals or social groups are, again, not restricted to the Nazi context. They were common, for example, to all eugenic research endeavors and policies that existed internationally since the late nineteenth and early twentieth centuries and that continued far beyond 1945.

The Nazi context, however, was necessary to bring the problematic potentials of all these widespread attitudes, motivations, and behavior patterns to a particularly radical manifestation. The Nazis promulgated a political program that claimed to be founded in biology and that enabled both political institutions and medical scientists to differentiate between human beings according to their supposed biological value. To be sure, this political program also existed in other national contexts, but there it was contested by rival political programs and alternative scientific theories and interpretations of heredity and the nature of man. The specificity of the Nazi context is first, and perhaps most importantly, rooted in the abolition of political and scientific pluralism in combination with a broad readiness of politicians, intellectuals, and the entire population and culture to follow the promises of the rationale and efficiency of science. As a consequence of the hegemonic and uncontested belief in the eugenic-racial hygienic paradigm's potential to differentiate human beings according to their supposed biological value, the human individual lost his place in the center of medical care.

NOTES

1. The last attempt for a full bibliography on the topic dates from 1995; it comprises more than 450 pages: Christoph Beck, *Sozialdarwinismus, Rassenhygiene, Zwangssterilisation und Vernichtung "lebensunwerten" Lebens. Eine Bibliographie zum Umgang mit behinderten Menschen im "Dritten Reich" und heute*, 2nd ed. (Bonn, 1995); on the development of the historiography of Nazi medicine, see Volker Roelcke, "Trauma or Responsibility? Memories and Historiographies of Nazi Psychiatry in Postwar Germany," in *Trauma and Memory: Reading, Healing, and Making Law*, ed. Austin Sarat, Nadav Davidovitch, and Michal Alberstein (Stanford, CA: Stanford University Press, 2007), 225–242.
2. Gerhard Baader and Ulrich Schulz (eds.), *Medizin und Nationalsozialismus. Tabuisierte Vergangenheit—ungebrochene Tradition?* (Berlin, 1980); Michael Kater, *Doctors under Hitler* (Chapel Hill: The University of North Carolina Press, 1985); Paul J. Weindling, *Health, Race, and German Politics between National Unification and Nazism, 1870–1945* (London: Cambridge University Press, 1989); Franz-Werner Kersting, *Anstaltsärzte zwischen Kaiserreich und Bundesrepublik. Das Beispiel Westfalen* (Paderborn, 1996); Winfried Süss, *Der "Volkskörper" im Krieg. Gesundheitspolitik, Gesundheitsverhältnisse und Krankenmord im nationalsozialistischen Deutschland 1939–1945* (Munich, 2003).
3. Ernst Klee, *"Euthanasie im NS-Staat. Die Vernichtung "lebensunwerten Lebens,"* (Frankfurt, 1983); Götz Aly, Karl Heinz Roth et al., eds., *Beiträge zur nationalsozialistischen Gesundheits- und Sozialpolitik*, vol. 1ff. (Berlin, 1985ff.); Hans-Walter Schmuhl, *Rassenhygiene, Nationalsozialismus, Euthanasie. Von der Verhütung zur Vernichtung "lebensunwerten Lebens" 1890–1945* (Göttingen, 1987); Michael Burleigh, *Death and Deliverance* (London: Cambridge University Press, 1994); Henry Friedlander, *The Origins of Nazi Genocide. From Euthanasia to the Final Solution* (Chapel Hill: University of North Carolina Press, 1995); Heinz Faulstich, *Hungersterben in der Psychiatrie 1914–1949. Mit einer Topographie der NS-Psychiatrie* (Freiburg/Br., 1998); Petra Fuchs, Maike Rotzoll, Ulrich Müller et al., eds., *"Das Vergessen der Vernichtung ist Teil der Vernichtung selbst": Lebensgeschichten von Opfern der nationalsozialistischen "Euthanasie"* (Göttingen, 2007).
4. See, e.g., Benno Müller-Hill, *Tödliche Wissenschaft. Die Aussonderung von Juden, Zigeunern und Geisteskranken 1933–1945* (Reinbek bei Hamburg, 1984); Götz Aly and Karl Heinz Roth, "Der saubere und der schmutzige Fortschritt," *Beiträge zur nationalsozialistischen Gesundheits- und Sozialpolitik* 2, (1985); Volker Roelcke, Gerrit Hohendorf, and Maike Rotzoll, "Psychiatric research and 'euthanasia.' The case of the psychiatric department at the University of Heidelberg, 1941–1945," *History of Psychiatry* 5 (1994): 517–532; Paul J. Weindling, *Epidemics and Genocide in Eastern Europe 1890–1945* (Oxford: Oxford University Press, 2000); Volker Roelcke, "Psychiatrische Wissenschaft im Kontext nationalsozialistischer Politik und 'Euthanasie.' Zur Rolle von Ernst Rüdin und der Deutschen Forschungsanstalt für Psychiatrie," in *Die Kaiser-Wilhelm-Gesellschaft im Nationalsozialismus. Bestandsaufnahme und Perspektiven der Forschung*, ed. Doris Kaufmann (Göttingen, 2000), 112–150; the case studies by Angelika Ebbinghaus, Karl-Heinz Roth, and Thomas Werther in *Vernichten und Heilen. Der Nürnberger Ärzteprozess und seine Folgen*, ed. Klaus Dörner and Angelika Ebbinghaus (Berlin, 2001); Hans-Walter Schmuhl, "Hirnforschung und Krankenmord: Das Kaiser-Wilhelm-Institut für Hirnforschung 1937–1945," *Vierteljahreshefte für Zeitgeschichte* 50 (2002): 559–609; Benoit Massin, "Mengele, die Zwillingsforschung und die 'Auschwitz-Dahlem Connection,'" in *Die Verbindung nach Auschwitz. Biowissenschaften und Menschenversuche an Kaiser-Wilhelm-Instituten,* ed. Carola Sachse (Göttingen, 2003), 201–254; Alexander von Schwerin, *Experimentalisierung des Menschen: Der Genetiker Hans Nachtsheim und die vergleichende Erbpathologie 1920–1945* (Göttingen, 2004); Florian Schmaltz, *Kampfstoff-Forschung im Nationalsozialismus. Zur*

Kooperation von Kaiser-Wilhelm-Instituten, Militär und Industrie (Göttingen, 2005); Hans-Walter Schmuhl, *Grenzüberschreitungen. Das Kaiser-Wilhelm-Institut für Anthropologie, menschliche Erblehre und Eugenik, 1927–1945* (Göttingen, 2005); and the case studies in Wolfgang U. Eckart ed., *Man, Medicine, and the State. The Human Body as an Object of Government-Sponsored Medical Research in the 20th Century* (Stuttgart, 2006).

5. For a rather isolated, early publication on the topic, see Natalija Decker, "Zur medizinischen Versorgung polnischer Zwangsarbeiter in Deutschland," in *Der Arzt als Gesundheitsführer. Ärztliches Wirken zwischen Ressourcenerschließung und humanitärer Hilfe im Zweiten Weltkrieg,* ed. Sabine Fahrenbach and Achim Thom (Frankfurt/Main, 1991), 99–107; for a general survey on the issues, see Andreas Frewer and Günther Siedbürger, eds., *Medizin und Zwangsarbeit im Nationalsozialismus. Einsatz und Behandlung von "Ausländern" im Gesundheitswesen* (Frankfurt/Main, 2004); Günther Siedbürger and Andreas Frewer, eds., *Zwangsarbeit und Gesundheitswesen im Zweiten Weltkrieg. Einsatz und Versorgung in Norddeutschland* (Hildesheim, 2006).
6. See, e.g., Hendrik van den Bussche, ed., *Medizinische Wissenschaft im "Dritten Reich". Kontinuität, Anpassung und Opposition an der Hamburger Medizinischen Fakultät* (Berlin, 1989); Wolfgang Woelk, Frank Sparing, Kerstin Griese, Michael G. Esch, eds., *Die Medizinische Akademie Düsseldorf im Nationalsozialismus* (Essen, 1997); Susanne Zimmermann, *Die Medizinische Fakultät der Universität Jena während der Zeit des Nationalsozialismus* (Berlin, 2000); Gerhard Aumüller, Kornelia Grundmann, Esther Krähwinkel et al., eds., *Die Marburger Medizinische Fakultät im "Dritten Reich"* (Munich, 2001); Bernd Grün, Hans-Georg Hofer, Karl-Heinz Leven, eds., *Medizin und Nationalsozialismus. Die Freiburger Medizinische Fakultät und das Klinikum in der Weimarer Republik und im "Dritten Reich"* (Frankfurt/Main, 2002); Ralf Forsbach, *Die Medizinische Fakultät der Universität Bonn im "Dritten Reich"* (Munich, 2006); Sigrid Oehler-Klein, ed., *Die Medizinische Fakultät der Universität Giessen im Nationalsozialismus und in der Nachkriegszeit* (Stuttgart, 2007).
7. Helga Jakobi, Peter Chroust, and Matthias Hamann, eds., *Aeskulap und Hakenkreuz. Zur Geschichte der Medizinischen Fakultät in Gießen zwischen 1933 und 1945* (Giessen, 1982; 2nd ed. Frankfurt/Main, 1989).
8. This connection between career opportunities and the content of medical research may be analyzed by the analytical concept of "career resources": see Volker Roelcke, "Funding the scientific foundations of race policies: Psychiatric genetics and the impact of career resources," in *Man, Medicine, and the State. The Human Body as an Object of Government Sponsored Medical Research in the 20th Century,* ed. Wolfgang U. Eckart (Stuttgart, 2006): 72–87.
9. For these developments and reconfigurations in psychiatry after 1933, see Roelcke, in *Man, Medicine, and the State,* 2006.
10. For this general analytical approach, see Volker Roelcke, "Wissenschaften zwischen Innovation und Entgrenzung: Biomedizinische Forschung an den Kaiser-Wilhelm-Instituten, 1911–1945," in *Sozialdarwinismus, Genetik und Euthanasie. Menschenbilder in der Psychiatrie,* ed. Martin Brüne and Theo Payk (Stuttgart, 2004), 92–109.
11. See Weindling, *Health, Race, and German Politics,* 1989; Süss, *Der "Volkskörper" im Krieg,* 2003.
12. For the consequences of the war on the redistribution of resources and its impact on health care policies and practices, see, e.g., Süss, *Der "Volkskörper" im Krieg,* 2003; Wolfgang U. Eckart and Alexander Neumann (eds.), *Medizin im Zweiten Weltkrieg* (Paderborn, 2006).
13. For this kind of atrocious research, see the literature listed in note 4 above, as well as Eckart, *Man, Medicine, and the State,* 2006, and Neumann, *Medizin im Zweiten Weltkrie,*

2006. For the perspective of the physicians in charge and, in particular, their explicit or implicit justifications of their activities, see Thomas Rütten, "Hitler with—or without—Hippocrates? The Hippocratic Oath during the Third Reich" in *Koroth* 12 (1996/97): 91–106; see also the chapters by Klaus Dörner and Ulf Schmidt in *Vernichten und Heilen. Der Nürnberger Ärzteprozess und seine Folgen*, ed. Klaus Dörner and Angelika Ebbinghaus (Berlin, 2001); and Paul Weindling, "'No mere murder trial': The discourse on human experiments at the Nuremberg Medical Trial" in *Twentieth-Century Ethics of Human Subjects Research. Historical Perspectives on Values, Practices, and Regulations*, ed. Volker Roelcke and Giovanni Maio (Stuttgart, 2004), 167–180.

14. See also Chapter 3 by William Seidelman in this book.
15. Kater, *Doctors under Hitler*, 1985. For more recent calculations and numbers, see Martin Rüther, "Geschichte der Medizin. Ärzte im Nationalsozialismus," *Deutsches Ärzteblatt* 98 (2001): C2561-C2562; as well as Forsbach, *Die Medizinische Fakultät der Universität Bonn im "Dritten Reich,"* 39–40.
16. Oehler-Klein, *Die Medizinische Fakultät der Universität Giessen im Nationalsozialismus und in der Nachkriegszeit*, 2007.
17. Kater, 1985, passim; Michael Kater, "Die soziale Lage der deutschen Ärzteschaft vor und nach 1933," in *Vernichten und Heilen. Der Nürnberger Ärzteprozess und seine Folgen*, ed. Klaus Dörner and Angelika Ebbinghaus (Berlin, 2001), 51–67; and Martin Rüther, "Ärztliches Standeswesen im Nationalsozialismus 1933–1945," in *Geschichte der deutschen Ärzteschaft*, ed. Robert Jütte (Köln, 1997), 143–193.
18. Astrid Ley, *Zwangssterilisation und Ärzteschaft: Hintergründe und Ziele ärztlichen Handelns 1934–1945* (Frankfurt/Main, 2003), 230–303.
19. Ricarda Stobäus, " 'Euthanasie' im Nationalsozialismus: Gottfried Ewald und der Protest gegen die, Aktion T4,' " in *"Euthanasie" und die aktuelle Sterbehilfe-Debatte. Die historischen Hintergründe medizinischer Ethik*, ed. Andreas Frewer and Clemens Eickhoff (Frankfurt/Main, 2000), 177–192; a further example for the refusal to meet the expectations of the regime is described by Alexander Neumann, "Die ärztliche Ethik hochgehalten—Der Militärarzt Dr. Christian Spiering im deutsch besetzten Norwegen" in *Zivilcourage. Empörte, Helfer und Retter aus Wehrmacht, Polizei und SS*, ed. Wolfram Wette (Frankfurt/Main, 2004), 241–255.
20. For a survey, see Paul Weindling, "International Eugenics: Swedish Sterilization in Context," *Scandinavian Journal of History* 24, 1999, 179–197.
21. Paul Weindling, "The Rockefeller Foundation and German Biomedical Sciences, 1920–1940: from Educational Philanthropy to International Science Policy," in *Science, Politics, and the Public Good*, ed. Nicolaas Rupke (Basingstoke/London, 1988), 119–140.
22. Volker Roelcke, "Programm und Praxis der psychiatrischen Genetik an der Deutschen Forschungsanstalt für Psychiatrie unter Ernst Rüdin: Zum Verhältnis von Wissenschaft, Politik und Rasse-Begriff vor und nach 1933," *Medizinhistorisches Journal* 37, 2002, 21–55; and Roelcke, in *Man, Medicine, and the State*, 2006.
23. See, e.g., Schmuhl, *Rassenhygiene, Nationalsozialismus, Euthanasie*, chapter 1.
24. Alexander Neumann, "Personelle Kontinuitäten—inhaltlicher Wandel: Deutsche Physiologen im Nationalsozialismus und in der Bundesrepublik Deutschland," *Medizinhistorisches Journal* 40, 2005, 169–189.
25. See Neumann, "Personelle Kontinuitäten—inhaltlicher Wandel," in particular 176–180; Karl Heinz Roth, "Flying bodies—enforcing states: German aviation medical research from 1925 to 1975, and the Deutsche Forschungsgemeinschaft," in *Man, Medicine, and the State. The Human Body as an Object of Government-Sponsored Medical Research in the 20th Century*, ed. Wolfgang U. Eckart (Stuttgart, 2006), 107–137.
26. Roelcke, in *Man, Medicine, and the State*, 2006.

27. See, e.g., the case studies by Marion Hulverscheidt, Gabriele Moser, Volker Roelcke, Karl-Heinz Roth and others in Eckart, *Man, Medicine, and the State*, 2006, as well as Anne Cottebrune, "Erbforscher im Kriegsdienst? Die Deutsche Forschungsgemeinschaft, der Reichsforschungsrat und die Umstellung der Erbforschungsförderung," *Medizinhistorisches Journal* 40, 2005, 141–168.
28. Angelika Ebbinghaus and Karl-Heinz Roth, "Kriegwunden. Die kriegschirurgischen Experimente," in *Vernichten und Heilen. Der Nürnberger Ärzteprozess und seine Folgen*, ed. Klaus Dörner and Angelika Ebbinghaus (Berlin, 2001), 177–218.
29. On aviation medicine, see Roth 2006; on medical genetics, see, e.g., Roelcke 2002 and 2006, Massin 2003; Schwerin 2004; Schmuhl 2005.
30. For these justifications, see the reconstructions listed in note 13.
31. Volker Roelcke, "Introduction" in *Twentieth-Century Ethics of Human Subjects Research. Historical Perspectives on Values, Practices, and Regulations*, ed. Volker Roelcke and Giovanni Maio (Stuttgart, 2004), 11–18.

CHAPTER 3

Academic Medicine during the Nazi Period: The Implications for Creating Awareness of Professional Responsibility Today

William Seidelman

In the mid-1930s, a young physician from Louisiana by the name of Michael DeBakey began a period of surgical training in Europe. After studying vascular surgery in France, he moved to Germany to continue his surgical studies with Professor Martin Kirschner at the University of Heidelberg.[1] Heidelberg was, at that time, recognized as a center for outstanding surgical training. Young DeBakey was following a decades old tradition whereby North American physicians traveled to the birthplace of modern scientific medicine—Europe.[2] It was the German-language universities that gave birth to many of the major discoveries and developments that have formed the foundation of modern medicine and medical science. DeBakey's European quest followed in the footsteps of such eminent physicians as the clinician William Osler, the pathologist William Welch, and the surgeon William Halsted. The period that saw an eruption in knowledge and technology in clinical and basic science also witnessed the parallel development of eugenics and scientific racism, which were also considered legitimate scholarly and scientific endeavors.

Eugenics was established by the British scientist Frances Galton.[3] The major application of eugenics—surgical sterilization of people with purported intellectual and mental disabilities—was first implemented in several states in the United States and in at least three provinces in Canada. The 1922 Model Sterilization Law, developed by the American eugenicist Harry Laughlin for the U.S. Congress, was at least as stringent as the 1933 Sterilization Law enacted by the Hitler government.[4] Eugenic sterilization in the United States was officially sanctioned by a 1927 decision of the Supreme Court.[5] In 1928 the Canadian province of Alberta passed a eugenic sterilization law that remained in effect until 1972. Ironically, the legislature of the province of British Columbia passed a eugenic sterilization law in 1933, the year Hitler came to power and enacted the German program of enforced eugenic sterilization. In the province of Ontario, hundreds of sterilizations were performed in the absence of any legislation.[6]

Analogous to the application of eugenics were the explicit and implicit policies of racial/religious/ethnic discrimination through such policies as racial segregation. Universities in the United States, Canada, and elsewhere instituted policies of racial selection and discrimination through the application of a *numerus clausus* (closed number) to restrict Jewish applicants to medical faculties. Many countries restricted the licensing of immigrant Jewish physicians who had been disenfranchised by the racist policies of Nazi Germany.[7] Ironically, in the 1920s and 1930s, Jewish medical students in the United States found it easier to gain admission to a medical school in Germany than in America. The late Dr. George Rosen, a noted American public health expert and historian, was one such person. After being rejected by American medical schools, Rosen obtained his medical degree in 1935 from the University of Berlin.[8]

While German scientists followed the eugenic practices of the United States and Canada with great interest, North American scientists and politicians were influenced by the scholarly activity of German experts. A singular example of the work of German academics was the textbook on human genetics, eugenics, and racial hygiene entitled *Menschliche Erblischkeitslehre*, which was authored by three leading German professors—Erwin Bauer, Eugen Fischer, and Fritz Lenz—and was first published in 1921 by J.F. Lehmanns Verlag of Munich. The publisher, Julius Lehmann, sent a copy of the 1923 second edition to Adolf Hitler, who was then incarcerated in Landsberg Prison. Hitler used the genetics text to substantiate the National Socialist ideology he had expounded in *Mein Kampf*, which he had written while at Landsberg. Professor Fritz Lenz prided himself on the fact that parts of the text were reflected in Hitler's speeches.[9] The textbook was immensely popular and published in five German-language editions. An English-language version of the third edition (1927) was published in 1931. It can still be found on the shelves of some libraries in North America.[10]

Universities and scientific research units of Germany rapidly incorporated the eugenic and racist policies and practices of the Hitler regime.[11] They were led by internationally noted professors and scientists. Jewish professors and scientists were dismissed, and ardent Nazis were appointed in their place. Jewish physicians were forbidden to treat non-Jews and lost their positions with hospitals and sick funds. Eugenics and racial hygiene were incorporated within the curriculum of the medical schools, including studies in eugenic/genetics clinics and laboratories. Student theses were written on eugenics subjects such as enforced sterilization.[12] After the outbreak of World War II and the large-scale killing programs, German medical schools received the cadavers of prisoners executed for political crimes or for trivial offenses, such as stealing a loaf of bread or a foreign laborer for socializing with a German woman.[13] In life, human beings were exploited for experiments that German law forbade being performed on animals.[14] In death, the bodies of victims of Nazi terror were exploited for collection by institutes of anatomy and pathology as well as for anthropological collections in museums.[15]

By 1938 and the Austrian surrender to Adolf Hitler, the academic and research establishment of Austria had the benefit of the five years experience of the Hitler regime in Germany. At the University of Vienna, the effort for racial cleansing was led by Eduard Pernkopf, the Austro-fascist and racist anatomy professor, who was the director of the Institute of Anatomy of the University of Vienna. Within days of the Nazi takeover, Pernkopf was appointed dean of medicine. A photograph of the Vienna faculty taken soon after the *Anschluss* (the political union of Austria with Nazi Germany in 1938) shows an amphitheater festooned with Nazi regalia; the dean, Professor Pernkopf, wearing Nazi insignia; and the entire assembled faculty standing with arms raised in a Hitler salute.[16]

The first racial science expert invited to Vienna was Professor Otmar von Verschuer, the German scientist and twin expert from the University of Frankfurt.[17] Verschuer was the mentor of Dr. Josef Mengele and eventually the principal investigator of at least some of the murderous Auschwitz twin experiments.[18]

Having had the benefit of the German experience, Pernkopf and his Vienna colleagues were able to complete the racist transformation of the University of Vienna much faster than the academic leaders in Germany.[19]

Professor Pernkopf became the director of a Vienna anatomy institute that had a proud tradition of pioneering leadership in human anatomy. His mentor was the renowned anatomist Ferdinand Hochstetter. Pernkopf, in partnership with the German medical publisher Urban and Schwarzenberg, undertook the development of a major atlas of human anatomy. The Pernkopf atlas was unique for three reasons:

1. It was created at a leading anatomical institute with one of the world's foremost medical faculties and published by Urban and Shwarzenberg, one of the world's leading publishers of medical texts.
2. Its watercolor paintings were of a very high caliber.
3. It was the first atlas of human anatomy to use the four-color offset method of printing. Together with the quality of the paintings, this printing technology resulted in illustrations that were unsurpassed in the excellence of their reproduction and could be reprinted on a large scale.[20]

The best example of the quality of the Pernkopf anatomical paintings was a painting by the artist Erich Lepier that portrays the neck dissection of a young man with a carefully trimmed moustache but coarsely shorn scalp.[21] It was this painting that raised questions in the mind of Professor Howard Israel while a member of the Faculty of Dentistry of Columbia University. Israel was puzzled by the relative youth of the subject and the manner in which the subject's hair had been crudely shorn. Israel asked himself if the subject could have been an executed prisoner.[22]

Professor Israel's suspicions were reinforced by the fact that some of the paintings incorporated Nazi insignia in the artists' signatures. The Holocaust Martyrs' and Heroes' Remembrance Authority, Yad Vashem, was approached and agreed to write to the Austrian authorities to request an investigation into the origins of the subjects portrayed in the atlas with specific reference as to whether they could have been victims of Nazi terror. The officials who initially responded on behalf of the university were not forthcoming with the truth.[23] The then rector of the University of Vienna, Professor Alfred Ebenbauer, realized that he was not being told the facts by some faculty in his university. In February 1997, the rector appointed an investigative committee headed by a historian outside the Faculty of Medicine.[24]

An interim report of the investigative committee was published in late 1997 with the final report released on October 1, 1998. The investigations documented that the Institute of Anatomy had received the cadavers of almost 1,400 persons executed in the Gestapo execution chamber at the Vienna regional court. The investigations also revealed the existence of almost 200 specimens from probable victims of Nazi terror at other institutes of the University of Vienna.[25]

Faced with the evidence, the University of Vienna acknowledged its role and undertook to retrieve every human specimen of unknown provenance. The collection of specimens was buried in the Jewish section of the Vienna Cemetery on Friday, March 22, 2002.[26]

Eight years passed between the time questions were first asked about the Pernkopf atlas in 1994 and the burial of the specimens in 2002. In the interim, the atlas had become a source of both local and international controversy. In response to the controversy, some universities removed the atlas from circulation; others included a special insert prepared by the University of Vienna concerning the origins of the atlas and the provenance of the specimens. In at least two instances, a special exhibit was prepared on the atlas and its origins.[27] Other universities did nothing. The Pernkopf atlas is no longer published. Some of the paintings were included in Clemente's *A Regional Atlas of Human Anatomy*. Professor Clemente, however, removed the Pernkopf paintings from the most recent edition.[28] The Pernkopf atlas can still be found in some libraries.

Despite the controversy associated with the anatomical specimens, there was an apparent absence of faculty and students from the University of Vienna at the burial ceremony. One would have expected a university presence that was more in keeping with the significance of the event and its importance to academic medicine beyond the university, the city of Vienna, and the Austrian nation.

The implications for academia are profound. The Vienna experience symbolizes the ongoing challenge for academic medicine, particularly in Germany and Austria where recognition and acknowledgment by the academic and professional communities of their role in the Third Reich have been problematic. That situation is beginning to change with some German and Austrian medical faculties sanctioning research and publication of their own institutional history during the Hitler period. German medical schools are also teaching the history of medicine during the Third Reich to undergraduates.[29]

As Professor Volker Roelcke said in his lecture at the Holocaust Museum Houston in 2007, there are three overriding issues arising from the tragedy of medicine in the Third Reich. First, there is the question of power; second, the hierarchy and priority of values; and third, the goal and impact of scientific and health care policy and programs that aim at the enhancement of human performance and the improvement of the genetic constitution of human society.

Within academic medicine, the levers of power are diverse and widespread among deans, administrators, clinical department heads, and researchers as well as external funding sources such as government and industry. The people who hold the power exert considerable influence over the institutions they lead and the people who work and study within those institutions. The hierarchy of values and their priority are represented by what gets taught, who gets what resources and for how long, and who gets appointments and tenure. The values are encompassed within the policies and programs expressed in the mission statement and strategic plans of institutions, funding bodies, and government. The allocation of resources by the institutional leadership determines how the mission and strategic initiatives are implemented.

In the end, what matters is how people receive care and how future health professionals are taught and trained to provide that care. The experience of Dr. and Mrs. DeBakey, attempting to get appropriate treatment for him in 2006, exemplifies the challenges faced by some people seeking care in our huge, complex systems—especially people who are elderly, handicapped, poor, frail, or socially marginalized.[30]

While it is easy to be critical, it should be emphasized that the vast majority of health care providers are outstanding both in terms of their clinical competence and their caring. Researchers are serious-minded individuals trying to improve the human condition. It is easy to highlight the negative and to be judgmental. In part, the medical profession may be its own worst enemy because it assumes or accepts increasing responsibility for complex issues ranging from global challenges such as AIDS (the next pandemic), conflict, and genocide to the application of the most recent discoveries be they in molecular biology and genetics or the application of material science to such things as joint replacement and drug-eluting coronary artery stents. The advancements and changes in every aspect of clinical medicine have been breathtaking.[31] It is remarkable what clinicians today are expected to know and to do, while at the same time addressing economic and political pressures, not to mention their own personal well-being and that of their loved ones. The public emphasis is on achievement and capacity, not the limits of that capacity or the vulnerability and fallibility of the health care providers and scientists who are merely mortal human beings.

It is difficult to address the subject of contemporary issues in medical ethics using the historical platform of medicine during the Nazi era.[32] It is understandable that people would be offended if it is inferred that a certain type of behavior is consistent with what happened during the Third Reich (i.e., something the Nazis did). This comprehensible aversion may explain why academic medicine has been reluctant to address this subject and why few medical faculties give any consideration to this subject in their curriculum. Like everyone else, physicians do not like to hear bad stories about themselves or their colleagues, particularly if there is an inferred implication that their behavior may in any way be associated with what happened during the Nazi era.

What is difficult to comprehend, but needs to be recognized, is the fact that the same faculties of medicine in Germany and Austria that gave us such stunning discoveries that revolutionized the world of medical science and continue to serve as the model for research and education also played a critical role in the implementation of the eugenic and racist policies and programs of the Nazi state. The universities and research institutes then and now had the power. Because of the incredible discoveries of the late nineteenth- and early twentieth-century governments, industry, and the universities and research institutes gave priority to scientific discovery and, in particular, to biomedical approaches with the goal of enhancing human performance and improving the genetic constitution of society. Today, we need to know more about how academic medicine and science evolved in the last century in Austria and Germany to better understand them and ourselves. We need to understand how a system and the people who comprised that advanced and sophisticated system could become party to evil. We need to know how it happened.

For many decades, the German and Austrian medical professions, universities, and research institutes resisted any examination of their role in the Third Reich. The Federal Chamber of Physicians of Germany claimed that most German doctors had been innocent bystanders and that the perpetrators convicted at Nuremberg were marginal and not representative of the mainstream.[33] We now know differently. The German and Austrian medical profession as well as academic and scientific institutions—such as the Max Planck research organization,[34] the University of Vienna, and the University of Giessen[35]—are beginning to address that problem. The truth is finally beginning to emerge, but there is much that we have yet to learn.

Notes

I wish to acknowledge the inspiration, leadership, and tireless effort of Dr. Sheldon Rubenfeld, who made the 2007 Michael DeBakey Medical Ethics Lecture Series at the Holocaust Museum Houston and this publication possible. I also wish to express my gratitude to Professor Volker Roelcke for his contribution to this chapter

1. Kirschner, a pioneer surgeon, performed the first successful pulmonary embolectomy. W. Bircks, "History of cardiac surgery in Germany—in consideration of her relation to the German Cardiac Society,"*Zeitschrift für Kardiologie* 91, no. 4 (2002): 81–85.
2. Harvey Cushing, *The Life of Sir William Osler*, 2 vols. (London: Oxford at the Clarendon Press, 1925); and Michael Bliss, *William Osler: A Life in Medicine* (Toronto: University of Toronto, 1999).
3. D.W. Forrest, *Francis Galton: The Life and Work of a Victorian Genius* (London: Paul Eleck, 1974).
4. Harry H. Laughlin, *Eugenical Sterilization in the United States* (Psychopathic Laboratory of the Municipal Court of Chicago, December 1922), http://www.people.fas.harvard.edu/~ wellerst/laughlin/.
5. Daniel Kevles, *In the Name of Eugenics: Genetics and the Uses of Human Heredity* (Berkeley and Los Angeles: University of California Press, 1985).
6. Angus McLaren, *Our Own Master Race: Eugenics in Canada, 1885–1945* (Toronto: McLellan & Stuart, 1990).
7. William E. Seidelman, "Medical Selection: Auschwitz Antecedents and Effluent," *Holocaust and Genocide Studies* 4, no. 4 (1989): 435–448.
8. Barbara Rosencrantz, "George Rosen, Historian of the Field," *American Journal of Public Health* 69, no. 2 (1979): 165–168.
9. Katrin Weigmann, "In the Name of Science," *EMBO Reports* 21, no. 10 (2001): 871–875.
10. Erwin Bauer, Eugen Fischer, and Fritz Lenz, *Human Heredity* (London: George Allan & Unwin Ltd., 1931).

11. Robert N. Proctor, *Racial Hygiene* (Cambridge: Harvard University Press, 1988); and Michael H. Kater, *Doctors Under Hitler* (Chapel Hill: University of North Carolina Press, 1989).
12. See Theresa M. Duello, "Misconception of 'Race' as a Biological Category: Then and Now" (lecture, Michael DeBakey Medical Ethics Lecture Series, Holocaust Museum Houston, 2007). See also chapter by same name in this book.
13. William E. Seidelman, "Medicine and Murder in the Third Reich," *Dimensions: A Journal of Holocaust Studies* 13, no. 1 (1999), http://www.adl.org/braun/dim_medicine_murder.asp.
14. William E. Seidelman, "Animal Experiments in Nazi Germany," *The Lancet* 1, no. 8491 (1986): 1214.
15. For example, Berichte, *Berichte: der Kommission zur Überprüfung der Präparatesammlungen in den medizinischen Einrichtungen der Universität Tübingen im Hinblick auf Opfer des Nationalsozialismus,* Herausgegeben vom Presidenten der Eberhard-Karls-Universität Tübingen Abdruck-auch auszugweise-nur mit Genehmigung des Herausgebers, 1990; Senatsprojekt der Universitäts Wien, *Untersuchungen Zur Anatomischen Wissenschaft in Wien: 1938–1945,* Wien, 1998; G. Aumüller and K. Grundmann, "Anatomy during the Third Reich: The Institute of Anatomy at the University of Marburg," *Annuals of Anatomy* 184 (2002): 295–303; and Christoph Redies et al., "Origin of corpses received by the anatomical institute at the University of Jena during the Nazi regime," *The Anatomical Record Part B: The New Anatomist* 285, no.1 (July 2005): 6–10.
16. See Eduard Pernkopf photograph, http://www.doew.at/service/ausstellung/1938/9/9_30.html.
17. Wolfgang Neugebauer, "Racial Hygiene in Vienna: 1938," Dokumentationsarchiv de öde mentationsar Widerstandes, wird publiziert in *wiener klinische wochenschrift,* Sonderheft, Marz 1998, http://www.doew.at/information/mitarbeiter/beitraege/rachyg.html.
18. Benno Müller-Hill, *Murderous Science: Elimination by Scientific Selection of Jews, Gypsies, and Others in Germany, 1933–1945* (Cold Spring Harbor, NY: Cold Spring Harbor Laboratory Press, 1998).
19. Gerald Weissmann, *They All Laughed at Christopher Columbus: Tales of Medicine and the Art of Discovery* (New York: Times Books, 1987).
20. David. J. Williams, "The History of Eduard Pernkopf's Topographische Anatomie des Menschen," *Journal of Biomedical Communication* 15 (Spring 1988): 2–12.
21. Helmut Ferner, *Eduard Pernkopf: Atlas of Topographical and Applied Anatomy,* 2 volumes (Philadelphia: W.B. Saunders, 1963), 234. See Fig. 25, des Menschennmie des Menschen Germany, 1933–1.
22. Howard A. Israel, "The Nazi Origins of Eduard Pernkopf's *Topographische Anatomie des Menschen:* The Biomedical Ethical Issues," *The Reference Librarian* 29, no. 61 and 62 (1998): 131–146.
23. William E. Seidelman, "From the Danube to the Spree: Deception Truth and Morality in Medicine" in *Jarbuch 1999,* ed. Ganglmair Siegwald, *Documentationarchiv des österreichischen Widerstandes* (Vienna: DÖW, 1999), 15–32.
24. University of Vienna, "Untersuchungen zur Anatomischen Wissenschaft an der Universitat Wien 1938–1945," press release, February 12, 1997.
25. Senatsprojekt der Universitäts Wien, 1998.
26. Bethany Bell, "Nazi Victims buried in Vienna," BBC News, March 22, 2002, http://news.bbc.co.uk/2/hi/europe/1888079.stm.
27. Medical Sciences Library, Ruth and Bruce Rappaport Faculty of Medicine, Technion: Israel Institute of Technology, Haifa, Israel, March, 2008; and *adobe medicus* bimonthly bulletin of the Health Sciences Library and Informatics Centre University of New Mexico

Health Sciences Center 31, no. 3, (May/June 2008): 7, http://hsc.unm.edu/library/adobemed/archive/adobe_medicus_v31no3_2008.pdf.
28. L. Gordon, "Tragic Beauty: The Story of Pernkopf's Anatomy" (lecture, Cedars-Sinai Emeritus Society, Cedars-Sinai Hospital, Los Angeles, CA, May 30, 2008.
29. Volker Roelcke, "Medicine During the Nazi Period: Historical Facts, and Some Implications for Teaching Medical Ethics and Professionalism" (lecture, Michael DeBakey Medical Ethics Lecture Series, Holocaust Museum Houston, 2007). See also chapter by same name in this book.
30. Lawrence K. Altman, "The Man on the Table Devised the Surgery," *New York Times*, December 25, 2006, http://www.nytimes.com/2006/12/25/health/25surgeon.html.
31. Jane Elliott, "A 60-year Revolution in Surgery," BBC News, June 28, 2008, http://news.bbc.co.uk/2/hi/health/7461004.stm.
32. Arthur L. Caplan, "Too Hard to Face," *Journal of the American Academy of Psychiatry and the Law* 33 (2005): 394–400.
33. William E. Seidelman, "Whither Nuremberg?: Medicine's Continuing Nazi Heritage," *Medicine and Global Survival* 2, no. 3 (September 1995): 148–157.
34. Speech given by Hubert Markl, President of the Max Planck Society for the Advancement of Science, on the occasion of the opening of the symposium entitled "Biomedical Sciences and Human Experimentation, Kaiser Wilhelm Institutes—The Auschwitz Connection," Berlin, June 7, 2001, quoted in *Medicine and Medical Ethics in Nazi Germany: Origins, Practices, Legacies*, ed. Francis R. Nicosia and Jonathan Huener (New York: Berghahn Books, 2002), 128–139.
35. Roelcke, "Medicine During the Nazi Period."

CHAPTER 4

MISCONCEPTIONS OF "RACE" AS A BIOLOGICAL CATEGORY: THEN AND NOW

THERESA M. DUELLO

The Third Reich's conception of a race was predicated on the belief that a "race"* was distinguished by a distinct group of genetically transmitted physical characteristics and that the inheritance of a small number of certain physical characteristics, such as white skin, blond hair, and blue eyes, was genetically linked to the inheritance of other traits deemed superior. The Nordic people bearing these characteristics were considered Aryans, or the "Master Race."

When Hitler and the Nazi party took power in Germany in 1933, they advocated eugenics, a social philosophy that proposed the improvement of the human species through interventions. This philosophy did not originate with Hitler; eugenic movements were well established in Great Britain and the United States at the time. Hitler, however, did develop his own version of eugenics, or "racial hygiene." Positive eugenics gave rise to efforts to encourage reproduction by those Aryans deemed to be genetically superior. Negative eugenics gave rise to efforts to prevent reproduction in order to cleanse future generations of the German people (race) of "inferior genes."

The Third Reich recognized that the Aryan "race" was not currently without defect and acted to prevent reproduction of those Aryans who exhibited traits believed to reflect the presence of "inferior genes." Abstinence and family planning were not advocated; voluntary or forced sterilization was. In 1933 the Third Reich enacted the German Law for the Prevention of Hereditarily Diseased Progeny (Sterilization Law), a law that required sterilization of congenitally diseased offspring to rid the German people of undesired traits.[1]

The law had three main provisions. First, any person suffering from a hereditary disease could be surgically sterilized if the experience of medical science showed that it was highly probable that the offspring would suffer from serious physical

* Quotation marks indicate incorrect or questionable use of a term.

or mental defect. Second, any person would be considered congenitally diseased if they suffered from one of eight hereditary diseases: hereditary "feeblemindedness," schizophrenia, manic-depressive insanity (also referred to as circular insanity), hereditary epilepsy, Huntington's chorea (also referred to as St. Vitus's dance), hereditary blindness, hereditary deafness, or severe hereditary bodily deformities. Third, any person suffering from severe alcoholism could be sterilized. The goal of the law was to achieve a 100 percent successful sterilization rate to remove undesired traits from future generations of the German people. It was considered the duty of medical professionals as a group to preserve and promote the nation's health, sound heredity, and racial purity.

In 1933 a Berlin correspondent for the *Journal of the American Medical Association (JAMA)* wrote, "Since sterilization is the only sure means of preventing the further hereditary transmission of mental disease and serious defects, this law must be regarded as an evidence of brotherly love and of watchfulness over the welfare of coming generations."[2]

The law required physicians to file a certificate with a hereditary health court to report any individual who suffered from a congenital defect or severe alcoholism. The hereditary health court, composed of a judge, a medical officer, and a medical practitioner, decided whether the individual should undergo sterilization, the use of force being permissible. It was stipulated that the physician reporting the congenital defect could not be the physician who provided the sterilization. The individual was then sterilized with or without his or her consent at the expense of the state.

That same year the Berlin correspondent for the *JAMA* wrote,

> The negative measures for the prevention of offspring with serious hereditary defects will be followed by positive legislative measures for the protection of families with sound hereditary attributes, and particularly the families with many children.. . .The personal desire to be sterilized, in itself, cannot and must not be considered important. The danger that person with such desires, born of the endeavor to secure unrestrained sexual gratification without fear of consequences, might find all too easily a willing surgeon, as well as other possible abuses, has been studiously avoided.[3]

Physicians were penalized for not performing required sterilizations as well as for providing unauthorized sterilizations to healthy Aryans for family planning purposes.

Blindness, deafness, epilepsy, or bodily deformities were physical traits apparent to a physician or a hereditary health court; their hereditary nature, however, had not been scientifically established and could not have been observable upon physical examination. Diagnosis of schizophrenia, manic-depressive insanity, Huntington's chorea, and severe alcoholism were within the scope of the diagnostic abilities of a well-trained physician as well. "Feeblemindedness" was not. Nowhere did the Sterilization Law or the hereditary health courts define "feeblemindedness." The Royal College of Physicians of London had previously defined a feebleminded individual as "one who is capable of earning a living under favorable circumstances, but is incapable, from mental defect existing from birth, or from an early age, of competing on equal terms with his normal fellows; or of managing himself and his affairs with ordinary prudence."[4] Thus, this category was not medically or scientifically defined. The inability to sustain oneself was attributed to genetics (i.e., nature rather than nurture) with the result that anyone deemed socioeconomically disadvantaged—the poor, the homeless, the uneducated, the unemployed—was potentially considered to be feebleminded. Absent a definition of feeblemindedness and absent an explanation of a diagnosis, the Third Reich, indeed, advanced a socioeconomic justification for the Sterilization Law, emphasizing the economic burden to the state posed by the care of these individuals, who were depicted as "useless eaters" or "life unworthy of life." Posters and figures in biology textbooks emphasized the cost of the upkeep of these individuals to the Third Reich, which bolstered the argument for sterilization of the "genetically inferior" who were economic burdens to the state. The Third Reich, therefore, sought a single, cost-effective approach for the successful surgical sterilization of large numbers of people, a method that could be applied uniformly by physicians at hospitals and clinics throughout Germany.

While I grew up well aware of the atrocities perpetrated in prison camps during World War II, I had no specific knowledge of the 1933 Sterilization Law or how the Sterilization Law segued into the euthanasia programs for the systematic mass murder of tens of thousands of those individuals deemed mentally and physically defective. Despite the fact that I hold a Ph.D. in human anatomy and was subsequently trained as a reproductive biologist, my early training was devoid of any formal education in medical history and medical ethics, even the subset that pertains to reproductive science. As an undergraduate student, I recall asking how the length of time it took a human fertilized egg to complete transit down the Fallopian tube to implant in the uterus was determined. I was told studies had been conducted in rabbits. I was not told that Professor Hermann Stieve of the University of Berlin and Berlin Charité Hospital had women impregnated and subsequently slaughtered at designated intervals to determine how long it took for the human egg to migrate.[5] Only when I switched my research focus from the rat reproductive system to the human reproductive system in the 1980s did I learn any history of medical ethics as a requirement for approval of a protocol to study human tissues. And even that history began with the Nuremberg Code, a set of principles for human experimentation formulated at the Doctors' Trial at the end of World War II. Not only had my own education been deficient throughout these years, but I had also failed to educate my students. I vowed to remedy this deficiency.

I began by reading *When Medicine Went Mad*,[6] *Doctors from Hell*,[7] and *Life Unworthy of Life*.[8] As I searched my university library's catalog of books, I discovered over 700 doctoral theses of German medical students archived in the University of Wisconsin Ebling Health Sciences Library. Six of these theses, written between 1937 and 1940, detailed the eugenic sterilization of hundreds of women deemed to be suffering from one of the "hereditary" disorders specified by the Sterilization Law.[9] I had them translated with the hope of gaining insight into this period of history through the eyes of the German medical students.

According to the Berlin correspondent writing for the *JAMA* in 1932, a course in medicine at this time from beginning through licensure took up to seven years: eleven semesters of study, one semester for examinations, and one year of internship.[10] Thus, these students matriculated in medical school in the early 1930s, when the theory of heredity and the study of eugenics became required coursework for students in grammar schools, high schools, and trade schools as well as medical schools throughout Germany and when the Sterilization Law became law.[11] In 1933 it was also decreed that a physician seeking admission to panel practice had to be enrolled in a federal register, which denied registration if the physician was not a German citizen in full standing, if the physician was of non-Aryan origin, or if the physician had a spouse of non-Aryan origin.[12] Medical students in the 1930s, therefore, were well aware of the emphasis on eugenics in the medical curriculum, of the role of their chosen profession in promoting racial hygiene, and of the requirement that German physicians in full standing be Aryan.

In the remainder of this chapter, I will first summarize the content of these doctoral theses, reflect on insights gleaned, and comment on their relevance to medical education today. I will then discuss the Third Reich's misconception of "race" (erroneously defined by skin color, eye color, and hair color) as a biological category and the persistence of incorrect definitions of "race" in medical research today.

The six translated theses, written by students at Albert-Ludwigs-Universität in Freiberg and Ruprecht-Karls-Universität in Heidelberg, detail the first five years of medical experience by the physicians at German regional hospitals with sterilizations of girls and women diagnosed with a hereditary disorder, principally "feeblemindedness." In all instances, the research was supervised and approved by physicians serving as clinic and hospital directors. These doctoral theses were similar in that they all provided extensive detail on the number of patients, the statement of the diagnosis, the type of surgical approach used, and the sterilization method used. They also provided extensive details regarding the number and type of postoperative complications as well as the cause of any deaths. While written as a requirement for the medical degree, they also contained detailed documentation of compliance with the Law for the Prevention of Hereditarily Diseased Progeny.

Particularly striking were the number and types of surgical sterilization approaches tested. A few will be cited as examples. The Menge method involved approaching the Fallopian tubes through the inguinal canal, resection of the tubes, and then fixing the tubal ends extraperitioneally in the inguinal canal. The Peitmann method for *tubenkeil* (tubal wedge/ungula) excision entailed removal of the isthmus of the Fallopian tube, stitching of the uterine wound, covering it with a peritoneal fold, and lowering of the distal tubal end below the peritoneum. The Madlener method entailed crushing the mucous membranes and musculature of the isthmal portion of the tube with pinchers to essentially cause scarring of the tubes and obliteration of the lumen of the tube. This method was commonly considered the least invasive, simple, safe, and reliable, but it could also result in the formation of peritoneal fistulas or adhesions to the ileum. To avoid these complications, physicians employed the Madlener-Walthard modification, which involved cutting the tube with the distal section being sewn under the peritoneum. Methods involving "high amputation of the cervix" were also cited, but this approach was given up because one-third of women subsequently became pregnant. While the term "experimental surgery" was never used, there is no question that the sheer number and variety of surgical approaches and methods indicated their experimental nature as physicians sought to identify a single method for uniform use.

The German medical students' theses differed in several significant respects from what would be considered a medical case study today. These differences provide insight into the impact of politics on the scientific climate as well as what was considered acceptable scholarship by the physicians serving as thesis advisors. I will focus on three areas that, for the sake of argument, I will refer to as aberrations.

First, the medical case studies presented in the doctoral theses proclaimed the virtue of the Sterilization Law for the improvement of the German people and cited the credit due to the Third Reich. Storch wrote, "The struggle of Germany to create congenitally sound future generations has and will continue its worldwide contribution to, totally or partly, preventing unsuitable elements from reproducing. Germany will strive, not only to keep the peace in Europe, but also to create a community of congenitally sound and strong people."[13] Strouvelle concurred stating, "The purpose of the law was to liberate the genotype of the German people from diseased genes to the greatest extent possible, to prohibit the danger of hereditary diseases for the future, and thus make them rarer with each passing generation."[14] He added, "It is solely due to the Third Reich that, for the very first time, legislation for the sterilization of congenitally diseased persons has been put in place for an entire nation."

At a time when the defeat in World War I was still fresh in the minds of the German people, racial hygiene provided a point around which physicians and medical students could rally. Koch alluded to the complexity of the issue, but agreed with Storch and Strouvelle when she wrote, "The question of sterilization for eugenic reasons comprises a complex array of related issues; even before the practical solution of a law preventing congenitally diseased children was reached, the issue has been thoroughly thought out and examined from a wide variety of viewpoints, including those regarding racial hygiene and hereditary biology, population politics and medicine, social and economic views, and last but not least, worldview."[15]

Objecting to what she believed to be unjust criticism by other nations, Storch wrote, "Although similar laws already existed in other countries, they have always tried to propagate against Germany, again and again. In 1909 in the state of Connecticut in the USA, a law of this kind was enacted for the first time in North America [*sic*]. That was long before a law on sterilization was accepted in Germany."[16]

Several theses also cited multiple problems with real or potential criticism of the law by the patients themselves. Bacher wrote, "At this point, it is too early to come up with any experience of long-term consequences. However, because they are possible and very much considered among laymen, the doctor has to especially take into account lamentations and complaints, which are brought up as a consequence of the sterilization operation. It is of fundamental importance to swiftly counter any public unrest concerning the measures taken due to the law, and to firmly establish the idea of hereditary health among our population."[17]

There were also allusions to problems resulting from the use of local anesthesia, which allowed patients to observe and subsequently talk about what had transpired; problems with the persistent objections of patients who were sterilized despite the fact their current pregnancy was allowed to go to term; and problems of housing sterilized patients with patients hospitalized for other reasons. Bacher cited both the objections of patients and the need to house them separately when he wrote, "A small number of those who the Court of Hereditary Health had decided should be sterilized had to be delivered to the hospital by police, a few in straightjackets. Fear of the operation is a more probable cause for this than aversion to the law. Some of the congenitally diseased behaved badly towards those around them during their stay at the hospital. It would be of great advantage to create a special ward if there is enough space. In this way, a great deal of talk of unauthorized persons about the operation and its possible consequences could also be prevented. Few of those sterilized keep quiet about the surgery they have been through."[18] The remarks of the students are amazingly frank and make one wonder whether the thesis advisors permitted or actually promoted such candor.

The second noticeable aberration in the theses was the lack of either the definition or justification of the diagnoses that led to the sterilization. The Sterilization Law required that physicians submit certificates to hereditary health courts to report a "congenitally diseased" individual. Presumably, the medical practitioner and possibly the medical officer advised the judge as to the individual's diagnosis. In these doctoral theses, the majority of the women and children were diagnosed as "feebleminded," but nowhere is there a single line, let alone a discussion, of what constituted feeblemindedness. Strouvelle's thesis lists the following distribution of hereditary disease in the 630 patients ranging in age from 13 to 45:[19]

Inborn Feeblemindedness	360
Schizophrenia	119
Manic Depressive Insanity	12
Hereditary Epilepsy	88
Inborn Blindness	16
Inborn Deafness	23
St. Vitus's Dance	1
Severe Alcoholism	2
Physical Deformities	9

He goes on to list the occupations of the women and children sterilized.

No occupation	508
Domestic servants	62
Workers	23
Seamstresses	13
Office workers	7
Sales clerks	8
School girls	6
Piano teacher	1
Apprentice nurse	1
Kindergarten teacher	1

Strouvelle also notes that the two largest groups—those with no occupation and domestic servants—were composed mostly of the innately feebleminded and schizophrenics.[20] This suggests that the working definition of the hereditary health courts was comparable to that of the Royal College of Physicians of London—a feebleminded individual was one "incapable, from mental defect existing from birth, or from and early age, (a) of competing on equal terms with his normal fellows; or (b) of managing himself and his affairs with ordinary prudence."[21]

Storch did concede a level of difficulty in assigning the diagnosis: "The vast difference when it comes to the number of feebleminded people probably has mostly to do with the difficulty in accounting for the actual number of feebleminded, because it is hard to draw the line between slight feeblemindedness and 'natural stupidity.' On the other hand, it is extremely important to come to grips especially with the slight degrees of feeblemindedness, because this degree is most likely to be the dominant inherited factor.... It becomes clear just how strong the hereditary factor of feeblemindedness is when you realize that when one parent is feebleminded, 40–50 percent of the children will be feebleminded, while this percentage increases to over 90 percent when both parents are feebleminded."[22]

Storch's description indicates that she viewed "feeblemindedness" as a single trait linked to a single specific gene. It is evident, however, that the hereditary health courts did deviate from their concept of Mendelian genetics in a few instances where it was decided sterilization was not required of an individual deemed a talented musician or a talented artist.

The medical students recognized that omission of any explanation or justification of the diagnosis that led to the sterilization left the doctors vulnerable to accusations that they violated the law, for which they might suffer consequences. Therefore, they provided detailed explanations and justifications of sterilizations that were either cancelled or delayed. For example, some procedures were deemed unnecessary because the reproductive tract had failed to develop normally or because the individual had been rendered sterile from a previous infection. Other procedures were delayed or cancelled because of acute fevers or complicating illnesses.

The third aberration was the representation of the law as a reflection of the legislators' concern for the safety and welfare of the patient when, in fact, the law dealt only with the goal of modifying the genetic makeup of future generations of the German people. Storch wrote, "We are bound by duty as physicians, and as the performing entity, to bestow the persons who are affected according to the law with the best care possible. The most important thing still today has to be to safeguard life. In second place comes the reliability of a successful result."[23]

While it appears that the emphasis was on the well-being of the patient, she goes on to state, "All leading persons and physicians in Germany are in agreement that no failure will harm the law."[24] Tietge states, "The best method of surgery must be chosen, which gives the patient the best chances of a quick release, accommodating the safety of the patient with regards to life, complications and success of the operation."[25] Strouvelle cites a physician who performed hysterectomies (rather than tubal sterilizations), claiming to have been led by the legislators' goal of achieving a 100 percent rate of success. Acknowledging that hysterectomy was associated with a 2–3 percent mortality rate, Strouvelle states this is in direct contrast to the view held by participants in the 1935 Conference of the German Gynecological Society, namely, that safety of life should be the leading principle in sterilizing operations.[26] All of the authors take this position, but Strouvelle identifies the 1935 conference as the source. Collectively, these statements reflect an attempt on the part of these doctoral students to reconcile the Sterilization Law with the Hippocratic Oath.

My study of these theses as primary resources has provided the opportunity for a number of lines of inquiry, not the least of which is medical education under the Third Reich. Do they explain the T4 Euthanasia Program or the Holocaust? No, but they provide insight into how a cohort of German physicians and the medical students they trained may have rationalized the Sterilization Law and the euthanasia programs to rid the German people of undesired traits. Today they serve as excellent case studies, encouraging students to develop personal medical ethics that will enable them to recognize and resist external threats to their patients' well-being.

I will use these doctoral theses as teaching tools in the health disparities courses I teach to undergraduate and medical students to underscore past and present fears of racial hygiene and genetic selection. Racial and ethnic minorities—as well as women, children, and the disabled—have historically been underrepresented in medical research. In 1993 the National Institutes of Health issued the "Guidelines for the Inclusion of Women and Minorities in Clinical Research" in an attempt to address this underrepresentation.[27] However, historical fear and distrust of medical research discourage participation in clinical studies and generate concern that genetic information could be used to limit access to health care, health insurance, and life insurance.

I will now discuss the Third Reich's misconception of "race"—defined as skin color, eye color, and hair color—as a biological category and the persistence of similar misconceptions of race in medical research today. It is impossible to know definitively how the Third Reich defined race. There is no single definition to which all scientists or laypeople ascribe today, so it is a highly improbable there was anything approximating a consensus then. While social scientists have long understood race to be a social construct of no biological significance, it has been less clear to biologists and physicians. The confusion in the United States stems in part from the use of social categories developed for federal program data collection to define populations in medical studies that examine biological, rather than social, endpoints. For example, from 1790 to 1850 the U.S. Census Bureau used only two racial categories—white and black—for purposes of data collection.[28] Subsequently, more categories were added: American Indian (1860), Asian and Pacific Islander (1910), Mexicans (1930), Spanish-speaking origins (1940), and mixed ancestry (1950). In 1977 the Office of Management and Budget Directive 15 attempted to clarify its intentions in the use of these racial categories (May 1977; revised October 1997):

> This Directive provides standard classifications for record keeping, collection, and presentation of data on race and ethnicity in Federal program administrative reporting and statistical activities. These classifications should not be interpreted as being scientific or anthropological in nature, nor should they be viewed as eligibility for participation in any Federal program. They have been developed in response to need expressed by both the executive branch and the Congress to provide for the collection and use of compatible, nonduplicated, exchangeable racial and ethnic data by Federal agencies.[29]

Nonetheless, categories of race and ethnicity are routinely used to describe patient or client populations in medical studies. As recently as 2002, the U.S. Patent Office approved a patent for the combination of two generic antihypertensive drugs (BiDil®) for the treatment of heart failure in African American patients. In 2005 the Food and Drug Administration approved the drug for treatment of "self-identified

African Americans," although there was no sound scientific data that demonstrated that hypertension is a unique disease entity in Americans identified on the basis of skin color, facial characteristics, and hair type.[30]

The physicians and biologists who have pondered over the use of race as a biological or genetic category are divided in their opinions. There are those who argue that race is a biological category and those who argue that race is a legitimate proxy for a biological category. Some recognize that defining race as a biological category is not ideal, but are resigned to its historical use in the medical literature. Others argue that the need to categorize a person by race will be obviated by the advent of individualized medicine, that is, medicine based on any individual's known genetic makeup. And there are still others, this author among them, who argue that race is not a biological category and that the seminal issue is genetic ancestry, not race, which is a social category.

Given this confusion and lack of consensus, it is relevant to ask to what extent the Nazi misconceptions of race persist today. In 1933 Nazis defined the "Master Race" by external physical characteristics—skin, hair color, hair type—that did not confirm the presence or absence of a genetic disease. Similarly, in 2005 African Americans were asked to "self-define" as a race by external physical characteristics—skin, hair color, hair type—that do not confirm the presence or absence of a genetic disease.

What information do we not have or are we not processing that should dispel this misconception today? We now know that race has no genetic basis. Phase I of the Human Genome Project demonstrated that as humans we are 99.9 percent identical at the DNA level. We know there is not one characteristic, trait, or gene—or even a small number of characteristics, traits, or genes—that distinguishes all members of one race from all members of another race. The 0.1 percent in which we vary does not correlate with the social designations of race. In fact, the Human Genome Project confirmed that there is more genetic variation *within* than *between* races.

To demonstrate that skin color is not a genetic classification in my courses, I draw on my training in human anatomy, specifically histology. Both "white" and "black" individuals possess the melanin gene that instructs certain cells, melanocytes, to produce the pigment melanin, which can be seen in microscopic sections of the skin. In an individual where the cell is actively transcribing the melanin gene, the result is the production of the black melanin protein, which gives the skin a dark appearance. In an individual where the gene is not actively transcribing the melanin gene, the result is the absence of the black melanin protein, which gives the skin a lighter appearance. Both individuals are *Homo sapiens sapiens.*

Kingdom: Animalia
  Phylum: Chordata
    Class: Mammalia
      Subclass: Eutheria
        Order: Primates
          Family: Hominidae
            Genus: Homo
              Species: sapiens
                Subspecies: sapiens

Both the human species and subspecies are "sapiens" to reflect the fact that subspecies of humans do not exist. Nothing in this classification or in microscopic sections of skin reveal anything about hair texture, eye color, blood type, artistic talent, musical talent, athletic ability, or the ability to excel at math.

In conclusion, this journey from realizing the inadequacies of my understanding of medical history through the discovery and study of these doctoral theses has completely reframed how I advise and mentor my premedical students. In the past, I emphasized taking upper-level biology courses, including genetics, in order for them to convince an admissions committee of their ability to make an easy transition to graduate-level coursework. I now strongly encourage students to also enroll in courses in anthropology, population biology, medical history, and medical ethics in order not to have the deficits in their education that I have come to realize were in my own. I also use the doctoral theses as teaching tools since there is no comparison between telling students simply that "human subjects violations have occurred" and reading that "it is hard to draw the line between slight feeblemindedness and 'natural stupidity.'"[31]

NOTES

I wish to express my gratitude to Dr. Sheldon Rubenfeld and Dr. William Seidelman for their support, their many helpful discussions, and especially their belief that a biologist was not displaced pursuing a topic in medical history. I would also like to thank Micaela Sullivan-Fowler, Greg Prickman, Mary Hitchcock, archivists at the University of Wisconsin Ebling Library, for their unending assistance; Dr. Erik Eriksson for translation of the theses; and Dr. Ray Purdy for hours of helpful discussions of the problems inherent in the use of "race" in the biological medical literature.

1. United States Holocaust Memorial Museum, "Victims of the Nazi Era, 1933–1945: Handicapped." United States Holocaust Memorial Museum, www.ushmm.org/education/resource/handic/handicapbklt.pdf, 3.
2. Correspondent in Berlin, "Sterilization to Improve the Race," *JAMA* 101(1933): 866–867.
3. Ibid.
4. H.H. Goddard, *Feeblemindedness: Its Causes and Consequences* (New York: Macmillan, 1914), 4.
5. Letter from Prof. Friedrich Vogel of Heidelberg to Prof. Jürgen Peiffer of Tübingen, September 12, 1997. Quoted by permission of Prof. Vogel in a letter from Prof. Peiffer to William Seidelman, December 3, 1997. Cited in William Seidelman, *The Legacy of Academic Medicine and Human Exploitation in the Third Reich* (Baltimore: John Hopkins University Press, 2000), 328.
6. Arthur Caplan, ed., *When Medicine Went Mad: Bioethics and the Holocaust* (Totawa, NJ: Humana Press, 1992).
7. James M. Glass, *Life Unworthy of Life: Racial Phobia and Mass Murder in Hitler's Germany* (New York City: Basic Books, 1997).
8. Vivien Spitz, *Doctors from Hell: The Horrific Account of Nazi Experiment on Humans* (Boulder: Sentient Publications, 2005).
9. The six translated doctoral theses are as follows: Erich Bacher, "Bericht über 210 weibliche Sterilisationen" (doctoral thesis, Ruprecht-Karls-Universität Heidelberg, originally

published in Würzburg by Buchdruckerei Richard Mayr, 1940), 1–13; Gertrud Koch, "Das Erlebnis der Unfruchtbarmachung bei weiblichen Erbkranken" (doctoral thesis, Albert-Ludwigs-Universität Freiberg Freiberg i. Br., 1937), 1–19; Elwine Schuhmacher, "Bericht über 480 weitere eugenische Tubensterilisierungen unter besondere Berücksichtigung der Frage der menstruellen Zyklusverschiebung nach Tubensterilisierungen" (doctoral thesis, Ruprecht-Karls-Universität Heidelberg, originally published in Würzburg by Buchdruckerei Richard Mayr, 1939), 1–23; Antonia Storch, "Mortalität und Morbidität bei eugenischen Sterilisierungen an 190 Frauen" (doctoral thesis, Ruprecht-Karls-Universität Heidelberg, completed in the Städt Hospital in Speyer in 1935, originally published in Speyer by Pilger—Druckerei GMBH-Verlag, 1939), 1–32; Karl Strouvelle, "Erfahrungen bei der Sterilisation weiblicher Erbkranker auf Grund von 630 Fällen des Landeskrankenhauses Homburg/Saar" (doctoral thesis, Ruprecht-Karls-Universität Heidelberg, originally published in Würzburg by Buchdruckerei Richard Mayr, 1939), 1–29; Wolfgang Teitge, "Die Erfahrungen mit Mengesterilisierungen an der Landesfrauenklinik Erfurt im Vergleich mit denen anderer Methoden" (doctoral thesis, Albert-Ludwigs-Universität Freiberg i. Br., 1938), 1–20.

10. Correspondent in Berlin, "Prospects of Medical Students in Germany," *JAMA* 99 (1932): 47–48.
11. Correspondent in Berlin, "Eugenics as a Subject for Medical Students," *JAMA* 98 (1932): 830; Corr. in Berlin, "New Regulations Governing Medical Students," *JAMA* 99 (1932): 666; and Correspondent in Berlin, "Eugenics as a New Subject for Medical Students," *JAMA* 101(1933): 722–723.
12. Correspondent in Berlin, "The Admission of Physicians to Panel Practice," *JAMA* 103 (1934): 501.
13. Storch, "Mortalität und Morbidität."
14. Strouvelle, "Erfahrungen bei der Sterilisation."
15. Koch, "Das Erlebnis der Unfruchtbarmachung."
16. Storch, "Mortalität und Morbidität." Indiana, not Connecticut, enacted the first sterilization act in North America in 1907.
17. Bacher, "Bericht weibliche Sterilisationen."
18. Ibid.
19. Strouvelle, "Erfahrungen bei der Sterilisation."
20. Ibid.
21. Goddard, *Feeblemindedness*, 4.
22. Storch, "Mortalität und Morbidität."
23. Ibid.
24. Ibid.
25. Teitge, "Die Erfahrungen mit Mengesterilisierungen."
26. Strouvelle, "Erfahrungen bei der Sterilisation."
27. U.S. Department of Health and Human Services, *NIH Guidelines on the Inclusion of Women and Children as Subjects in Clinical Research-Updated August 2, 2000*, Office of Extramural Research, http://grants.nih.gov/grants/guide/notice-files/NOT-OD-00-048.html.
28. Campbell Gibson and Kay Jung, *Historical Census Statistics on Population Totals By Race, 1790 to 1990, and By Hispanic Origin, 1970 to 1990, For The United States, Regions, Divisions, and States*, U.S. Census Bureau, September 2002, http://www.census.gov/population/www/documentation/twps0056/twps0056.html
29. Executive Office of the President of the United States, *Standards for the Classification of Federal Data on Race and Ethnicity*, Office of Management and Budget, June 9, 1994, http://www.whitehouse.gov/omb/fedreg/notice_15.html.

30. See Joseph Graves Jr., *The Emperor's New Clothes: Biological Theories of Race at the Millennium* (Piscataway, NJ: Rutgers University Press, 2001) or Joseph Graves Jr., *The Race Myth: Why We Pretend Race Exists in America* (New York City: Penguin Books, 2004) for a summary of common errors in the association of race and complex behavior.
31. Storch, "Mortalität und Morbidität."

CHAPTER 5

MAD, BAD, OR EVIL: HOW PHYSICIAN HEALERS TURN TO TORTURE AND MURDER

MICHAEL A. GRODIN

INTRODUCTION AND BACKGROUND

This work is part of a 25-year project on the problem of evil.[1] From work on Holocaust survivors[2] and contemporary torture survivors,[3] I turned my focus to bystanders[4] and then finally to perpetrators, particularly focusing on the role of physicians in torture.[5] The research I compiled is more than just an academic interest or a set of historical facts. The problem of evil is not a philosophic one; rather, it is a problem that continues to confront the world in new and ever varying forms. Though there are few instances in recorded history of massive and atrocious perpetration of evil that parallel the Holocaust, the information gained from studying the Nazis can be applied to the unfortunately many occurrences of torture and human rights abuses today. By studying the Holocaust and the people who carried it out, we are not only standing as witnesses to this terrible tragedy, but learning how to actively prevent such evil from occurring again.

NUREMBERG

The history of torture as a crime against humanity began in Nuremberg in 1945. After the International Military Tribunal, or "Big Trial," of military commanders,[6] 12 subsequent trials were conducted by the U.S. government using U.S. law, under the auspices of the Allies. Each of these trials was against a different category of defendants, such as German industrialists[7] and judges.[8] The first of these trials was the *U.S. v. Karl Brandt et al.*, the so-called Doctors' Trial[9] (see figure 5.1).

The Doctors' Trial was a murder trial, unusual in that all 23 defendants were physicians and scientists on trial for war crimes and crimes against humanity. The

Figure 5.1 The Defendants' Box at the Doctors' Trial.
Source: Courtesy of the United States Holocaust Memorial Museum (USHMM).

chief counsel for the prosecution, Brigadier General Telford Taylor, described the significance of the trial in his opening statement:

> The defendants in this case are charged with murders, tortures, and other atrocities committed in the name of medical science. The victims of these crimes are numbered in hundreds of thousands . . . It is our deep obligation to all peoples of the world to show why and how these things happened. It is incumbent upon us to set forth with conspicuous clarity the ideas and motives which moved these defendants to treat their fellow men as less than beasts. The perverse thoughts and distorted concepts which brought about these savageries are not dead. They cannot be killed by force of arms. They must be cut out and exposed, for the reason so well stated by Mr. Justice Jackson in this courtroom a year ago—"the wrongs which we seek to condemn and punish have been so calculated, so malignant, and so devastating, that civilization cannot tolerate their being ignored because it cannot survive their being repeated." (Military Tribunal 1, Case 1, 1946–1947.)

Racial Hygiene, Murder, and Genocide

Taylor understood the need to recognize and actively denounce the evils of the Holocaust and implored the international community to take a stand against evil. In the foreword to *The Nazi Doctors and the Nuremberg Code: Human Rights in Human Experimentation*, Professor Elie Wiesel asked, "How was it that physicians could have been involved in such atrocities?"[10] One might well ask how any human being could have been involved. But what Wiesel and Taylor recognized was that physicians have a special moral standing in their communities and in society at large—by nature of their advanced education and their oath to serve and protect humanity, physicians have voluntarily undertaken a special responsibility. What were the personal, professional, and political contexts that allowed physicians to use their skills to torture and kill rather than to help and heal? Some insight into the events of the Holocaust in the historical accounts of the role of medicine and physicians in relation to racial hygiene theories, the medicalization of social ills, and the meshing of medicine with nationalist socialist ideology may provide answers.

The idea of racial hygiene emerged at the turn of the twentieth century, and the racial policies of the Third Reich were in many ways adapted from eugenics practices developed in the United States in the early twentieth century.[11] Before the National Socialist Party came to power in Germany, there were already several institutes of racial hygiene at various German universities. The theories at these institutes grew out of the "science of eugenics" employed in the United States to justify the 23 separate state laws that allowed for the involuntary sterilization of individuals being supported by the government.[12]

Ultimately, the Nazis would carry this ideology beyond sterilization. They not only eliminated "undesirables" from their society, but also developed multiple programs for the creation of a "master race," including the *Lebensborn* program, which encouraged members of the SS to have children with women who had Aryan traits.[13] All the while they highlighted the "therapeutic" facet of their programs, claiming that destroying the unworthy was "purely a healing treatment."[14]

Eugenics was only one of many facets of the biological front the National Socialists put on their policies. Nazi leaders considered their political philosophy to be "applied biology" and adopted many public health policies in addition to those guided by social Darwinism, including antitobacco initiatives.[15] They medicalized their political movement and often referred to Hitler as the "great doctor of the German people."[16] Perhaps attracted by the medical metaphors, doctors flocked to the cause of National Socialism.

Joining a political party is one thing; using its ideology to justify the torture and extermination of an entire people is quite another. To see why physicians and scientists took that extreme step, we must examine the perpetrators within a framework of individual and group psychology, as well as in the larger social context.

PSYCHOLOGY OF THE INDIVIDUAL PERPETRATOR

First, it is impossible to explain the acts of torturers and murderers without understanding something of the psychology of human behavior, including the concepts of self-deception, the unconscious, drive, defense, aggression, narcissism, a permissive superego, and social service for an ideal. These psychological ideas are rooted in philosophical theories about human nature. Before examining an individual psyche, it is important to consider the view one has of humanity. There is a fundamental tension between classic and romantic visions of human reality that is highly relevant to an examination of perpetrators of torture. In the "classic" view, we all intrinsically have the capacity to do evil and are very precariously constrained by order and tradition; in other words, we all have the potential to be torturers. In the "romantic" view, men and women are intrinsically good, but are spoiled by circumstance and culture—this vision of human reality is full of possibilities currently constrained by society. Under this dual framework, individuals either have the capacity for evil and are prevented from acting by socialization and social constraints or are moral beings turned into torturers by evil social contexts. But the truth of human psychology is probably not so extreme.

DEHUMANIZATION

There are several psychological mechanisms by which individuals can overcome the social conditioning that prevents them from becoming perpetrators of atrocities. Dehumanization is a key psychoanalytic defense mechanism that allows individuals to avoid fully processing troubling events. Dehumanization of the self and of others draws on other defense mechanisms, including unconscious denial, repression, depersonalization, isolation of affect, and compartmentalization (the elimination of meaning by disconnecting related mental elements and walling them off from each other).[17] Ultimately, dehumanization allows the perpetrator to go beyond hatred and anger and commit atrocious acts as if they were part of everyday life.

There are two types of dehumanization processes. First, there is self-directed dehumanization, a diminution of an individual's own sense of humanness and self-image, which is often seen in cases of complex posttraumatic stress disorder (CPTSD). For torture survivors or other persons exposed to extended trauma, this process is a form of self-protection.[18] The second type is object-directed dehumanization, where others are perceived to lack human attributes. The two processes are mutually reinforcing, as reducing the self adds to reducing the object, and reducing the object adds to reducing the self. Perpetrators accomplish the dehumanizing process by making the other (the object) dirty, filthy, and physically less than human. One could argue that there is an increased ability to dehumanize others today secondary to the advent of technology, as modern warfare, automation, urbanization, specialization, bureaucratization, and the mass media all contribute to the isolation and objectification of individuals. Anonymity and impersonality cause a fragmented sense of one's role in society, contributing to dehumanization. Sometimes, dehumanization can be adaptive; for example, in a crisis, dehumanization of the injured or sick allows for an efficient rescue. Certain occupations classically teach and perhaps require selective dehumanization, including law enforcement and military and medical professions. This allows professionals to detach from full emotional responsiveness in the moment, but it also can be very dangerous.

SPLITTING

Dehumanization by itself cannot completely explain the healing-killing paradox. Splitting as a model of personality enables people to deal with trauma.[19] This is a form of self-deception in which the unconscious mind can wall off the conflict to eliminate incompatibilities with self-image, separating thought and even actual events from feeling. For a perpetrator, splitting can be used to rationalize and justify his actions, and, through reaction formation, he can convince himself that he is doing good or even that he is a hero. Robert Lifton's interviews of Nazi physicians and their surviving families revealed how far splitting or "doubling" (as Lifton terms it) went for those individuals. The Nazi physicians split the self: they saw themselves as healers with special powers, practically omnipotent, and killing became a part of healing—in their minds, one had to kill the enemy to heal one's people, one's military unit, and one's self.[20] Under this mental paradigm, there is no paradox in using Red Cross trucks to carry victims to a death camp or in wearing white coats while systematically killing children for experimentation: medicine becomes the equivalent of war, and physicians medicalize and humanize killing even while they dehumanize the victim.

NUMBING

Splitting is combined with numbing to more effectively distance the perpetrator from the victim. Psychic numbing diminishes the capacity to feel. Blocking feelings leads to extreme repression, including denial to the point of disavowal of what one perceives and derealization to the point that the victims never existed in the perpetrators' consciousness. One Polish survivor who worked in a medical block of a concentration camp partly defended Polish doctors who mistreated Jewish inmates by noting that "people grow indifferent to certain things. Like the doctor who cuts up a dead body [to do a postmortem examination] develops a certain resistance."[21] However, this numbing process was not completely successful as many physicians selecting at the ramps still needed to self-medicate with heavy drinking.[22]

OMNIPOTENCE

Concentration camp officials' omnipotent control over life and death was balanced by the Nazis' vision of themselves as one important part of a larger omnipotent social machine. The medical profession is susceptible to feelings of omnipotence, and Holocaust survivor Bruno Bettelheim suggests that "it is this pride in professional skill and knowledge, irrespective of moral implications" that makes physicians vulnerable to becoming perpetrators.[23] Ultimately, however, doctors are impotent to control death and disease, and this is part of the death anxiety that many physicians, especially officials, have. For the doctors in the Nazi party, omnipotence merged with sadism—they took pleasure in domination and control—but they still needed to eradicate their own vulnerability and susceptibility to pain and death; there is a powerlessness associated with omnipotence. They merged their anxiety over powerlessness into their pride at being part of the German war machine. The Nazi party was able to manipulate particular psychological vulnerabilities in individuals as it pressed them into serving the wishes of the group. In the mental struggle to maintain their professional identity, the Nazi physicians saw Hitler as the "father physician" and unified as a group beneath him (see figure 5.2).

Figure 5.2 Adolf Hitler as the Physician to the German People.
Source: Courtesy of USHMM.

MEDICALIZATION

The Nazi doctors medicalized, technicalized, and professionalized their activities; for example, telling themselves that a doctor's task is to alleviate suffering, they would use medical and technical skills to diminish the pain of the victims while setting up mass murder. They became absorbed in the technical aspects of medical work, examining inmates as a criterion for sending them to the gas chambers.[24] They became robotic actors, and the process of murdering became a performance. In their medical uniforms, acting as the Nazi male ideal, they used their professional power to ward off their death anxiety, killing to hold back death. Medical professionals have a special capacity to split: while an individual is part of the healing profession, everything he does must be healing. Through these justifications and within the larger social context of "political medicine," the Nazi doctors were able to mentally transform murderous acts into acts of healing.

PSYCHOLOGY OF GROUPS OF PERPETRATORS

In identifying themselves as part of a larger machine working to "heal" society, Nazi physicians diffused responsibility to the group rather than taking individual responsibility for their actions. They achieved group unity through the creation of special group language—euphemisms for the evil acts they carried out. They saw themselves as part of a "special" group, elite and important. There was a certain sense of belonging and being part of a movement. This group unity was facilitated by specialized training, ritualization of their actions, and, as previously discussed, the self-directed dehumanization and splitting, which allowed them to subsume individual identity while acting in a professional capacity.

Two key psychology experiments after World War II examined obedience to authority and diffusion of responsibility in groups, and further demonstrated the ease with which previously well-adjusted individuals can engage in evil activities.

OBEDIENCE TO AUTHORITY

Beginning in 1961, Dr. Stanley Milgram performed a set of experiments at Yale University in which subjects were asked to "deliver electro-shocks" to another person. When the experimenter told them to do so, 65 percent of the subjects used what they believed were dangerously high level of shocks.[25] In a later experiment, one-third of subjects continued the shocks when they were close enough to touch the person being shocked.[26] The key to these experiments is that someone else—an authority figure—accepted responsibility for the final outcome. Milgram postulated three categories of reasons for obeying or disobeying authority: first, a personal history of family or school background that encouraged obedience or defiance (i.e., learned object relations); second, a feeling of comfort derived from obeying authority, which is known as "binding"; and third, the sense of discomfort people get when they disobey authority.[27] All of the test subjects believed the experimenter was responsible for any consequences and presumed the legitimacy of the experiment.

When considering the effect that group dynamics can have on individuals, it is important to note what draws certain persons toward certain groups. Authority-oriented persons have a preference for hierarchy and clearly demarcated power relationships—they enjoy obeying and giving orders.[28] Such persons highly value obedience, and, if self-guidance is impossible, they will seek external guidance, joining groups such as the military to provide an opportunity for external orders and to fill inner emptiness. Interviews with the widows of SS officers reveal that several such men reported a "need to belong."[29] Authority-seeking persons also avoid confrontation with authority figures (such as strict and abusive parents), instead seeking to attain closeness with them in order to feel secure. These individuals may be even more likely to respond to authority than the average people who acted as subjects in Milgram's experiments.

DIFFUSION OF RESPONSIBILITY

Ten years after Milgram's landmark work, Dr. Philip Zimbardo simulated prison life among college students in the famous Prison Experiment at Stanford University, randomly assigning housemates to be either a guard or a prisoner. Within six days, the subjects had changed from university students who were friends and roommates to abusive controlling guards and servile prisoners.[30] Prisoner students became passive, dependent, and helpless. Guards expressed feelings of power and group belonging. They placed all responsibility for their actions on the researchers and the group as a whole, rather than accepting blame for individual actions. The experiment became violent and had to be ended early. Dr. Zimbardo, who had acted as prison superintendent as well as the principal investigator, concluded that the experiment demonstrated the ways in which situational factors can cause inhumane behavior, in this way corresponding with the Milgram experiments.

The manner in which the subjects of these experiments placed all responsibility on the shoulders of the principal investigators and/or the group parallels the manner in which the perpetrators of the Holocaust denied the possibility of being blamed when they had merely been following orders. Hitler often stated of his military conquests that he took the responsibility upon himself and, in doing so, provided the basis for his subordinates to psychologically exempt themselves from moral standards or judgment.[31] This diffusion of responsibility can occur in any situation of mass violence, whether hierarchically structured or not.

THEORIES OF AGGRESSION

Some theories of aggression focus on individuals as perpetrators, particularly on the idea that the desire to inflict violence on others is a condition and/or expression of primary sexual drive.[32] This idea focuses on sadists for whom inflicting violence is sexually exciting and whose aggression is in the service of Eros. Sadism is in all people, but in some it splits off from regulating factors and becomes a dominant urge—in these people, there is a competitive wish for dominance over others. The satisfaction that comes from winning or from dominating another person becomes an uncontrollable urge in sadists. Similarly, sociopaths lack control over their urge to hurt others. As a result, sadists and sociopaths don't function effectively in the systematic infliction of violence through torture or genocide; they tend toward killing or hurting individuals. Sadism, as such, is not a sufficient explanation for the behavior of perpetrators of torture and mass violence.

Group behavior tends to rely on diminishing the conscious individual personality, focusing thoughts and feelings in a common direction, and giving emotion and the unconscious dominance over reason and judgment. As a result, ordinary persons whose urges are more easily subsumed than those of sadists or sociopaths are more effective killers, especially in a hierarchically structured setting such as the military. Interviews with a particular group of perpetrators, the Nazi Party elite *Schutzstaffel*, or German SS, showed that they were not psychopaths but ordinary men.[33] As Hannah Arendt suggested in her work on the Eichmann trial, the evil perpetrated by Eichmann and the SS was not a function of deeply rooted malevolence, but merely a lack of imaginative capacity and a result of not thinking out the impact of their actions.[34] Her idea of the "banality of evil," though much criticized as downplaying the significance of traumatic acts of violence, captures the ease with which some evil acts are perpetrated. For example, it was not difficult to get doctors to kill 100,000 German mental patients. Given the sheer number of people required as active participants or at least complicit in that "euthanasia" program, it is highly unlikely that all the doctors involved were deviants. Instead, the fragmentation and division of labor allowed each individual to excuse their participation by saying that they "only" did their particular assigned tasks.

UNIQUENESS OF GROUP

Why do people participate in torture? Theories of obedience and diffusion of responsibility explain how individuals may be drawn into groups that perpetrate evil, like torture, but it is harder to understand how these groups initiate torture in the first place. One might assume that the purpose of torturers is to elicit information or an admission of guilt; to intimidate; to justify repression or revenge, real or perceived; to establish superiority; or to elevate themselves, but that does not address the psychology of the group that creates and facilitates situations of torture. Groups such as the Nazis used oaths of loyalty to bind each individual and used rituals to create a mystical atmosphere that further drew members in and separated them from the outside. When a group has a shared mystique and common values, the members develop camaraderie, a devotion to the organizational ideology and cause, and a sense that they are part of the elite. They find pride in performing difficult and important acts, and become completely subordinated to the organization. After a certain level of indoctrination, it becomes difficult to deviate from or defy the group. This binding prevents individual members from resisting participation in torture.

USEFULNESS OF TRAINING

Beyond the binding to a group, individuals often receive special training to mold them into torturers (see table 5.1). The indoctrination and training of a torturer often includes abuse. In the Nazi regime, members of the SS were carefully selected, beginning with individuals who were comfortable obeying authority, often because of a personal history (family or school background) that encouraged obedience. Starting with that foundation, groups are able to shape torturers through a series of steps. First, members are screened for intellect, physical ability, and a powerful positive identification with the political regime.[35] This not only helps groups find individuals with the abilities they want, but also fosters an idea among members that inclusion is special and the group is elite, differentiating members from others. New members are bound to the group through basic training, a set of initiation rites that often includes isolation from people outside the group, and the imposition of new rules and values. From this beginning, members develop an elitist attitude and an in-group language. The members learn to dehumanize themselves as well as outsiders—to

Table 5.1 The Formation of a Torturer

Select for personal history of obeying authority
Screen for intellect, physical strength, and positive identification with politics
Bind with initiation rites, isolation, new rules, new values
Use elitist language
Dehumanize and blame
Harass, intimidate, desensitize, promote instinctive responses
Reward obedience
Demonstrate social modeling of group violence
Regularize and routinize violence
Practice controlled violence

subsume their individual identities within that of the group. Leaders harass and intimidate recruits, preventing logical thinking and instilling instinctive responses.[36] Rewards are given for obedience, and socialization of the group includes witnessing group violence, often in the form of intimidation of recalcitrant members.[37] As a result, members become desensitized to violence; both seeing and perpetrating violence become routine. All of this training adds up to complete control of the group over its members.

PHYSICIAN VULNERABILITY

With this understanding of group psychology, it is easy to see how members of the military are susceptible to becoming perpetrators. It may be less obvious why medical doctors are vulnerable (see table 5.2). One must consider that physicians are experts at compartmentalization, who deal with life and death every day and whose profession carries a sense of power. The motivation for choosing a career as a physician is often a fantasy of power, either sadistic or voyeuristic, as medicine gives license to look, touch, and control. Doctors medicalize and dehumanize their patients so that they can more easily process what they have to do and deal with the suffering to which they are daily exposed—using science to objectify their work, they heal by attacking and killing disease with surgery or therapy or whatever tools they have available. Medical students also go through an initiation ordeal. In anatomy class, they handle a dehumanized cadaver or watch operations without knowing the patients and are made to feel shame for any lapses in which they show too much "weakness" or inability to dehumanize patients.[38] Medicine as a profession contains the rudiments of evil, and some of the most humane acts of medicine are only small steps away from real evil. For example, though a surgery to amputate a gangrenous limb is a healing act, it involves cutting and maiming of the human body, which under nonmedical circumstances would be an act of harm and criminality.

During the Holocaust, the paradigm of the Nazi Doctor was Josef Mengele. In the concentration camps, Mengele often assumed a dual role with his victims—acting like a parent, playing games, and giving sweets before brutally killing children in his twin experiments. He exhibited signs of obsessive-compulsive disorder, fixating on cleanliness and perfection in his experiments even when the patients he treated would shortly be consigned to their death.[39] In his 21 months at Auschwitz, Mengele performed elaborate research on twin children—probing, infecting, cutting,

Table 5.2 Why Physicians are Vulnerable to Becoming Perpetrators

Compartmentalization
Tendencies toward sadism, voyeurism
Healing through hurting, repressing awareness of violence
Use of science to objectify violence
Use of metaphors and euphemisms
Justifying and rationalizing
Medicalizing
Narcissistic grandiosity

and exposing them to painful procedures and, ultimately, murder with no anesthetic. One of his assistants, Miklos Nyiszli, described the experiments:

> In the work room next to the dissecting room fourteen Gypsy twins were waiting and crying bitterly. Dr. Mengele didn't say a single word to us and prepared a 10 cc and 5 cc syringe. From a box he took Evipal and from another box he took chloroform, which was in 20 cc glass containers, and put these on the operating table. After the first twin was brought in . . . a fourteen year old girl . . . Dr. Mengele ordered me to undress the girl and put her head on the dissecting table. Then he injected the Evipal into her right arm intravenously. After the child had fallen asleep he felt for the left ventricle of the heart and injected 10 cc of chloroform. After one little twitch the child was dead, whereupon Dr. Mengele had her taken into the corpse chamber. In this manner all fourteen twins were killed during the night.[40]

One of the victims of Mengele's twin experiments offered a more personal account:

> It wasn't because his face was terrifying. His face could look very pleasant. But the atmosphere in the barracks before he came and the preparation by the supervisors was creating that atmosphere of terror and horror that Mengele was coming. So everybody had to stand still. He would, for example, notice on one of the bunk beds that a twin was dead. He would yell and scream, "What happened? How is it possible that this twin died?" But of course, I understand it today. An experiment had been spoiled.[41]

Mengele, though notorious, was not the only Nazi doctor who could dissociate the deaths he caused and the deaths that merely occurred "by accident" in the camps. SS doctors would kill and then have a meal, flog and then dress for dinner, torture and then listen to the opera and return to the camps. They used euphemisms to disavow the violence and dissociate their feelings—what they did was "medical ramp duty"; they "evacuated," "transferred," and "resettled" Europe's Jewish population.[42] With this special language, killing was no longer killing; it was a routine bureaucratic action.

Some types of doctors may be more or less predisposed to dehumanize patients—surgeons, for example, whose main interaction with their patients is violent and occurs while the patient is unconscious. But performing a healing function, psychic numbing, diffusion of responsibility, derealization, and compartmentalization—which occur within many different sectors of the medical profession—all lead to decreased feeling. Thus doctors, regardless of specialty, have the potential to become perpetrators, and in Nazi Germany, many did.

THE CULTURAL AND SOCIAL CONTEXTS THAT FACILITATE PERPETRATORS

The Nazi party ideology was portrayed in a medicalized way that attracted doctors. Writing in *Mein Kampf* on the German state, Hitler said, "Anyone who wants to cure this era, which is inwardly sick and rotten, must first of all summon up the courage

Figure 5.3 The poster reads, "You are sharing the load! A genetically ill individual costs approximately 50,000 Reichsmarks by the Age of Sixty."

Source: Courtesy of USHMM.

to make clear the cause of this disease."[43] In this "scientific" metaphor, the ultimate victims of the Nazi government were a threat—they posed a danger of contagion that could "infect" the German body politic, and without "purification," would pollute race and class. In this imagery, doctors were placed in the role of shaman—to treat not individuals but rather the group, becoming "physicians to the *volk* [people]."[44] The white-coated doctor became the black-robed priest, a professional capable of leading the biological soldiers on a mission of medical purification, eradicating the impaired and incurable.

In July 1933, the German government passed the Law for the Prevention of Hereditarily Diseased Offspring.[45] This sterilization program targeted mental disorders such as schizophrenia, manic-depressive disorder, and alcoholism along with inheritable physical diseases. In 1939 the elimination of disabled persons began with German politicians and doctors using the term "euthanasia" to describe the killings.[46] Psychiatrists and psychoanalysts played a major role in the killing of as many as 100,000 mentally and physically disabled persons between 1939 and 1941 in a project named Action T4, short for *Tiergartenstrasse* 4, which was the address of the Foundation for Welfare and Institutional Care (see figure 5.3).[47]

This medicalized and political "solution" to mental disorders and disability may have played a role in drawing psychiatrists and psychoanalysts into the regime. The Third Reich is often portrayed as decrying psychoanalysis; the Nazi party ceremoniously burnt the works of Freud along with the works of Marx and other "Jewish" thinkers who were seen as threatening the National Socialist state.[48] Despite this, some analysts remained in Germany to become a part of the Göring Institute. Those who stayed changed their ideas to mesh with the ideology of the ruling party, ultimately playing a large role in getting rid of "untreatable patients." Science was bent to the service of the Nazi party, and the new guiding spirit of Nazified psychoanalysis was employed to develop mental health treatments that aligned with the Third Reich's racist ideology.[49]

Many different social contexts combine to create a situation in which any person may become a torturer. In Nazi Germany, the integration of medical-scientific and political ideologies as well as economic pressures and social concerns about "race" made it easier for individuals to dehumanize their fellow citizens. The fervent nationalism and overwhelming support for the Third Reich made it difficult for people to lodge rational protests against the extermination of other human beings. In any situation in which human beings are divided into groups—the genetically pure versus the weak, the citizens versus the foreigners, the wealthy versus the poor—the oppression and discrimination against the non-favored group are facilitated because of the development of an "us and them" dynamic.

Medicine Betrayed—An International Problem

Although the atrocities perpetrated by the Nazis may be the most prominent human rights abuses in the global consciousness, torture and other inhumane acts are still carried out. Torture is practiced in over 150 countries.[50] In many countries, there is documented evidence of physician involvement, and torture can be particularly destructive when healers are involved. For example, a study in Turkey revealed that the regular occurrence of torture often involved coercing physicians into hiding their findings of abuse and harassing medical professionals who tried to resist.[51]

One victim of torture in Argentina, Jacobo Timmerman, reported his experience with physician-perpetrators:

> [H]e took my arm and very smoothly said "you know Jacobo that we doctors have many secrets . . . you see here . . . this blue is one of your arteries and I can inject here. You

> know that we have some substances that make you talk but always so painful because it affects your brain . . . so why can't you just talk and we can be friends." His presence was a symbol that a scientific instrument is with you when you are torturers.[52]

All forms of torture undermine the victim's sense of security and self-worth, but physician involvement more fully shatters the victim's trust. Physicians may be involved before, during, or after torture and may serve many separate roles: to supervise, to observe, to assist, to falsify medical records, and sometimes to treat a patient so that the torture can continue.

CONCLUSION

Regardless of the scale at which torture is perpetrated, there is an obligation incumbent on every human being to make all possible efforts to prevent and speak out against torture. As Martin Niemöller so famously said in the wake of the Holocaust,

> First they came for the socialists, and I did not speak out—because I was not a socialist. Then they came for the trade unionists, and I did not speak out—because I was not a trade unionist. Then they came for the Jews, and I did not speak out—because I was not a Jew. Then they came for me—and there was no one left to speak for me.[53]

We cannot give up hope even in the face of what seems to be widespread and never-ending human rights abuses. When the rabbis of the Talmud were asked if it was our job to repair and fix the world, they answered that "it is not upon you to complete the task, but neither are you free to desist from it" (Pirkei Avot 2:21).

NOTES

I would like to thank the Boston Psychoanalytic Society and Institute Julius Silberger Scholar Fund and the United States Holocaust Memorial Museum Research Institute Fund for financial support of this project [on the problem of evil]. I would like to thank the annual American Psychoanalytic Association workshop on Psychoanalytically Informed Care of Survivors of Torture and Refugee Trauma; James Frosch, M.D.; William Malamud, M.D.; Bennett Simon, M.D.; Roberta Apfel, M.D.; and Stephen Bernstein, M.D., for scholarly interchange and important criticism. I thank my colleagues George Annas, J.D., M.P.H., and Leonard Glantz, J.D., for 25 years of collaboration and for being role models of integrity and impeccable scholarship. I thank Rabbi Joseph Polak for three decades of friendship and for his insights and support of this project. Finally, I would like to thank my research assistant Denali Kerr for her expert scholarship and editorial assistance.

1. Parts of this work were discussed and published as "Physicians and Torture: Lessons from the Nazi Doctors," *International Review of the Red Cross* 89, no. 867 (September 2007).
2. George A. Annas and Michael A. Grodin, eds., *The Nazi Doctors and the Nuremberg Code: Human Rights in Human Experimentation* (New York: Oxford University Press, 1992), 17–31.
3. Linda Piwowarczyk, Alejandro Moreno, and Michael Grodin, "Health Care of Torture Survivors," *Journal of the American Medical Association* 284, no. 5 (2000); 539–541.

4. Michael Grodin, "Coming to the Rescue of None of My Business: The Effect of Responding vs. Ignoring on Community Health" and (Erich Lindemann Memorial Lecture, Harvard Medical School, Cambridge, MA, March 2005).
5. Michael Grodin, "Mad, Bad, or Evil: How Physician Healers Turn to Torture and Murder" (Julius Silberger Lecture, Boston Psychoanalytic Institute, Boston, MA, March 2000).
6. *Trial of German Major War Criminals by the International Military Tribunal* (London: H.M. Attorney-General, His Majesty's Stationery Office, 1946), *United States, United Kingdom, France and the Union of Socialist Soviet Republics v. Hermann Göring et al.*, 1945–1946, Vol. 1–22.
7. *Trials of War Criminals before the Nuremberg Military Tribunals under Control Council Law 10* (Washington, D.C.: Superintendent of Documents, U.S. Government Printing Office, 1952), Military Tribunal 1, Case 5, *United States v. Friedrich Flick et al.*, 1947, Vol. VI, 1–1268.
8. *Trials of War Criminals before the Nuremberg Military Tribunals under Control Council Law 10* (Washington, D.C.: Superintendent of Documents, U.S. Government Printing Office, 1951), Military Tribunal 1, Case 3, *United States v. Josef Altstoetter et al.*, 1947, Vol. III, 1–1236.
9. *Trials of War Criminals before the Nuremberg Military Tribunals under Control Council Law 10* (Washington, D.C.: Superintendent of Documents, U.S. Government Printing Office, 1950), Military Tribunal 1, Case 1, *United States v. Karl Brandt et al.*, 1946–1947, Vol. I, 1–1004; Vol. II, 1–352 (both hereafter cited in text as Military Tribunal 1, Case 1, 1946–1947).
10. Elie Wiesel, "Foreword," in *The Nazi Doctors and the Nuremberg Code: Human Rights in Human Experimentation,* ed. George A. Annas and Michael A. Grodin, (New York: Oxford University Press, 1992), vii–ix.
11. Stefan Kühl, *The Nazi Connection: Eugenics, American Racism, and German National Socialism* (Oxford, UK: Oxford University Press, 1994).
12. André N. Sofair and Lauris C. Kaldjian, "Eugenic Sterilization and a Qualified Nazi Analogy: The United States and Germany, 1930–1945," *Annals of Internal Medicine* 132 (2000): 312–319.
13. Mark Landler, "Results of Secret Nazi Breeding Program: Ordinary Folks," *New York Times,* November 7, 2006.
14. Robert Lifton, *The Nazi Doctors: Medical Killing and the Psychology of Genocide* (New York: Basic Books, 1986).
15. Robert N. Proctor, "The Anti-tobacco Campaign of the Nazis: A Little Known Aspect of Public Health in Germany, 1933–1945," *British Medical Journal* 313 (1996), 1450–1453.
16. Wolfgang Weyers, *Death of Medicine in Nazi Germany: Dermatology and Dermatopathology under the Swastika* (Philadelphia: Ardor Scribendi, Ltd., 1998).
17. Viola Bernard, Perry Ottenberg, and Fritz Redl, "Dehumanization: a composite psychological defense in relation to modern war," in *Behavioral Science and Human Survival,* ed. Milton Schwebel (Palo Alto, CA: Behavioral Science Press, 1965).
18. Stanley W. Jackson, "Aspects of Culture in Psychoanalytic Theory and Practice," *Journal of the American Psychoanalytic Association* 16 (1968): 651–670.
19. Lifton, *The Nazi Doctors.*
20. Ibid., 431.
21. Ibid., 240.
22. Lifton, *The Nazi Doctors,* and James Waller, *Becoming Evil: How Ordinary People Commit Genocide and Mass Killing* (New York: Oxford University Press, 2002).
23. Bruno Bettelheim, "Foreword" to Miklos Nyiszli, *Auschwitz: A Doctor's Eyewitness Account* (New York: Arcade Publishing, 1993), v–xviii.

24. Lifton, *The Nazi Doctors.*
25. Stanley Milgram, "Behavioral Study of Obedience," *Journal of Abnormal and Social Psychology* 67 (1963): 371–378.
26. Waller, *Becoming Evil.*
27. Stanley Milgram, *Obedience to Authority: An Experimental View* (New York: Harper Collins, 1973).
28. Waller, *Becoming Evil.*
29. Tom Segev, *Soldiers of Evil: The Commandants of Nazi Concentration Camps* (New York: McGraw-Hill, 1988).
30. Craig Haney, W. Curtis Banks, and Philip G. Zimbardo, "Interpersonal Dynamics in a Simulated Prison," *International Journal of Criminology and Penology* 1 (1973): 69–97.
31. Fritz Redl, "The Superego in Uniform," in *Sanctions for Evil*, ed. Nevitt Sanford and Craig Comstock (San Francisco: Jossey-Bass Inc, 1971).
32. Debra Kaminer and Dan J. Stein, "Sadistic Personality Disorder in Perpetrators of Human Rights Abuses: A South African Case Study," *Journal of Personality Disorders* 15, no. 6 (2001): 475–486.
33. Lifton, *The Nazi Doctors*, 14.
34. Hannah Arendt, *Eichmann in Jerusalem: A Report on the Banality of Evil* (New York: Penguin Books, 1963).
35. Mika Haritos-Fatouros, "The Official Torturer," in *The Politics of Pain: Torturers and Their Masters*, ed. Ronald D. Crelinsten and Alex P. Schmid (Boulder, CO: Westview Press, 1994).
36. Lifton, *The Nazi Doctors.*
37. Janice T. Gibson and Mika Haritos-Fatouros, "The Education of a Torturer," *Psychology Today* (November 1986): 50–58; and Ronald D. Crelinsten, "In Their Own Words," in *The Politics of Pain: Torturers and Their Masters*, ed. Ronald D. Crelinsten and Alex P. Schmid (Boulder, CO: Westview Press, 1994).
38. Heidi Lempp and Clive Seale, "The Hidden Curriculum in Undergraduate Medical Education: Qualitative Study of Medical Students' Perceptions of Teaching," *British Medical Journal* 329 (2004): 770–773.
39. Olga Lengyel, *Five Chimneys* (Chicago: Ziff Davis, 1947).
40. Miklos Nyiszli, *Auschwitz: A Doctor's Eyewitness Account*, trans. Richard Seaver and Tibere Kramer (New York: Arcade Publishing, 1993).
41. Eva M. Kor and Mary Wright, *Echoes from Auschwitz: Dr. Mengele's Twins: The Story of Eva and Miriam Mozes* (Terra Haute, IN: Candles Press, 1995).
42. Waller, *Becoming Evil.*
43. Adolf Hitler, *Mein Kampf* (Boston: Houghton Mifflin, 1943, originally published 1925–1926), 435.
44. Weyers, *Death of Medicine in Nazi Germany.*
45. Robert N. Proctor, "Nazi Doctors, Racial Medicine, and Human Experimentation," in *The Nazi Doctors and the Nuremberg Code: Human Rights in Human Experimentation*, ed. George A. Annas and Michael A. Grodin (New York: Oxford University Press, 1992), 17–31.
46. Michael Burleigh, *Ethics and Extermination: Reflections on Nazi Genocide* (Cambridge, UK: Cambridge University Press, 1997).
47. Hans-Georg Güse and Norbert Schmacke, "Psychiatry and the Origins of Nazism," *International Journal of Health Services* 10, no. 2 (1980): 177–196; and Johannes Meyer-Lindenberg, "The Holocaust and German Psychiatry," *British Journal of Psychiatry* 159 (1991): 7–12.

48. Rose Spiegel, Gerard Chrzanowski, and Arthur H. Feiner, "On Psychoanalysis in the Third Reich," *Contemporary Psychoanalysis* 11, no. 4 (1975): 477–510.
49. Geoffrey Cocks, *Psychotherapy in the Third Reich: The Göring Institute*, 2nd ed. (New Brunswick, CT: Transaction Publishers, 1997).
50. Amnesty International, Annual Report 2006: The State of the World's Human Rights, Library, http://www.amnesty.org/en/library/info/IOR61/015/2006/en.
51. Physicians for Human Rights, *Torture in Turkey and Its Unwilling Accomplices* (Cambridge, MA: Physicians for Human Rights, 1996).
52. Jacobo Timmerman, *Prisoner Without a Name, Cell Without a Number* (Harmondsworth, UK: Penguin Books, 1982).
53. Martin Niemöller, "First They Came" (speech circa 1946, discussed in 1976 interview), http://www.history.ucsb.edu/faculty/marcuse/niem.htm.

CHAPTER 6

GENETIC DIVERSITY HAS PREVAILED, NOT THE MASTER RACE

FERID MURAD

Genetic heterogeneity and diversity have prevailed as primitive tribes migrated throughout the world and adapted to their environment. The Nazi plans for the "Master Super Race" were obviously flawed from the start, scientifically as well as morally. We have learned from history and science that inbred families have a greater incidence of genetic disorders. This has been well documented with some of the royal families in the world with marriages between family members. Today we can witness a similar phenomenon with Pima Indians in Southern Arizona. With sedentary life style, excess calories and alcohol, and intratribal marriages, the incidence of a complex genetic disorder, diabetes mellitus, can be as great as 50–70 percent in some families. There are undoubtedly many other examples.

Personally, I believe that the diverse groups of immigrants to the United States in the past 200 years have resulted in a remarkably heterogeneous genetic pool. This heterogeneity and the resulting mixed marriages between individuals from diverse countries and genetic backgrounds and the work ethic they have inherited have been major factors in the productivity, strength, and creativity of this country.

As an example, since 1901 there have been a total of 743 Nobel Laureates in Chemistry, Physics, Medicine, Peace, and Literature and 58 in Economics. Since World War II, more than half of the Laureates in Science (Physics, Chemistry, and Medicine) have been U.S. citizens or residents. Even more interesting, many of these have been immigrants or offspring of Jewish families from Eastern Europe who relocated because of the Holocaust.

Hitler's Plan for the Master Super Race was terribly flawed from the start. It resulted in the unfortunate death of millions. Furthermore, Germany and Eastern Europe lost some of their most talented people to the United States and other countries to compliment the gene pool elsewhere.

Part 2

Medicine after the Holocaust

CHAPTER 7

GENETICS AND EUGENICS: A PERSONAL ODYSSEY

JAMES D. WATSON

The science of genetics arose to study the transmission of physical characteristics from parents to their offspring. When closely studied, much variation exists for virtually any characteristic, say, in size or color, among the members of all species, be they flies, dogs, or ourselves, the members of the *Homo sapiens* species. The origin of this variability long fascinated the scientific world, which already in the nineteenth century asked how much of this variation is due to environmental causes (nurture) as opposed to innate hereditary factors (nature) that pass unchanged from parents to offspring. That such innate heredity exists could never be realistically debated. One need just look at how characteristics in the shape of the face pass through families. Ascribing, say, the uniqueness of the Windsor face to nurture as opposed to nature goes beyond the realm of credibility.

The key conceptual breakthrough in understanding the nature component of variation came in the mid-1860s from the experiments of the Austrian monk and plant breeder Gregor Mendel (1822–1884). In his monastery gardens, he created, by self-breeding, strains of peas that bred true for a given character like pea color or pod shape. Then he crossed his inbred strains with each other and observed how the various traits assorted in the progeny pea plants. In his seminal scientific paper, published in 1865, Mendel showed that the origin of this hereditary variability lay in differences in discrete factors (genes) that pass unchanged from one plant generation to another.

Genes first were of interest because they were the source of the variability between the members of a species, but they soon began to be appreciated more properly as the source of information that gives an organism its unique form and function. Its collection of genes (its genome) is what gives each organism its own unique developmental pathway. A dog is a dog, a bacterium a bacterium, et cetera, because of the information carried by their respective genomes. Gene duplication prior to cell division, thus, must be based on a very accurate copying process. Otherwise, there would be no constancy of genetic information and of the development processes they make possible. Correspondingly, genetic variation arises when genes are not accurately copied (mutated) and give rise to changed (mutant) genes.

As soon as the first spontaneous gene mutations became known, they were perceived as the obvious source of the new genetic variants necessary for Darwinian Evolution by "survival of the fittest." Many more dysfunctional than functional genes, however, likely result from random mistakes in the gene copying process. Thus, the rate at which the gene copying process itself makes mistakes will also be under strong evolutionary pressure. If too many spontaneous mutations occur, none of the mutant-gene-bearing organisms are likely to develop and produce viable offspring. Correspondingly, too low a mutation rate will not generate sufficient gene variants to allow species to compete effectively with those species evolving faster because of their more frequent generation of biologically fitter offspring.

Eugenic Solutions for Human Betterment

The coming together of Darwinian and Mendelian thinking immediately raised the question of the applicability of the new science of genetics to human life. To what extent was human success due to the presence in their recipients of good genes that led to useful traits like good health, social dependability, and high intelligence? Correspondingly, how many individuals at the bottom of the human success totem pole were there because they possessed gene variants perhaps useful for earlier stages in human evolution but now inadequate for modern urbanized life? Social Darwinian reasoning viewed the sociocultural advances marking humans' ascent from the apes as the result of continual intergroup and interpersonal strife, with such competitive situations invariably selecting for the survival of humans of ever-increasing capabilities. Social Darwinism came naturally to the moneyed products of the industrial revolution, a most prominent one being the talented statistician Francis Galton (1822–1911). Early in his career, he wrote the 1869 treatise "Hereditary Genius," later coining the term "eugenics" (from the Greek meaning well born) for studies that would bring about improvements of the human race through the careful selection of parents.

The eugenics movement naturally became galvanized by the new laws of Mendelian heredity. But immediately, the eugenicists' hopes had to be tempered by the fact that human genetics never would have the power of other forms of genetics where genetic crosses could be made as well as observed. For better or worse, the eugenicists' main research tool had to be hopefully well-collected, multigenerational pedigrees of physical and mental traits that passed through families from one generation to the next.

Initially, there were hopes that simple Mendelian ratios would characterize the inheritance of a broad-ranging group of human traits. But, in addition to the limitations brought about through the inability to confirm genetic hypotheses through genetic crosses, many of the studied traits appeared in too few families for appropriate statistical analysis. Particularly difficult to analyze were progeny traits not present in either parent. Conceivably, individuals had inherited one copy of the same recessive gene from each parent. Such tentative conclusions became more convincing when the respective traits, like albinism, were found more often in highly inbred, isolated populations where marriages of cousins were frequent.

Easier to assign as bona fide genetic determinants were dominant-acting genes that need be inherited from only one parent for their presence to be felt. Once Mendelian thinking had appeared, the inheritance mode of Huntington's disease, the terrible neurological disease that leads to movement and cognition disorders, was quickly ascertained as a dominant gene disorder.

Important as Huntington's disease was to the individuals and families of those so afflicted, the main focus of early twentieth-century eugenicists soon moved to potential genetic causations for disabilities of the mind, embracing a wide spectrum of manifestations from insanity through mental defectiveness, alcoholism, and criminality to immorality. With poorhouses, orphanages, jails, and mental asylums all too long prominent features of the most civilized societies, eugenicists with virtually religious fervor wanted to prevent more such personal and societal tragedies in the future. They also desired to reduce the financial burdens incumbent on civilized society's need to take care of individuals unable to look after themselves. But in their evangelical assertions that genetic causations lay behind a wide variety of human mental dysfunctions, the early eugenically focused geneticists practiced sloppy, if not downright bad, science and increasingly worried their more rigorous geneticist colleagues.

American Eugenics: Sloppy Genetics for the Legitimation of Class Stratification

The most notable American biologist, whose conclusions went far beyond his facts, was Charles B. Davenport (1866–1944), who parlayed his position as director of the Genetics Laboratory at Cold Spring Harbor, New York, to establish in 1910 a Eugenics Record Office using monies provided by the widow of the railway magnate E.H. Harriman. In his 1911 book, *Heredity in Relation to Eugenics*, pedigrees were illustrated for a wide-ranging group of putative hereditary afflictions ranging from bona fide genetic diseases, such as Huntington's disease and hemophilia, to behavioral traits of much less certain hereditary attribution, such as artistic ability and mechanical ability with reference to shipbuilding.[1] With, however, so little then known about the functioning of the human brain, Davenport's early rush to associate highly specific accomplishments of the human brain to specific genetic determinants could not automatically be dismissed as nonsense.

In addition to its family pedigree assembly and archival roles, the Cold Spring Harbor Eugenics Record Office frequently counseled individuals with family backgrounds of genetic diseases, particularly when they were considering marriage to blood relatives. Many such seekers of help must have been misled by advice that never should have been given, considering that era's limited power for meaningful genetic analysis. Worries about insanity were a major concern, where manic-depressive disease was seen to move through some families as if it were a dominant trait. In contrast, schizophrenia often had aspects of a recessive disease that appeared without warning and was not seen in the next generation.

If causative genes were indeed recessive, siblings of individuals displaying mental instability were at high risk of being carriers of insanity-provoking genes silently passing through their respective family trees. Marriage most certainly should be avoided between related individuals having severe mental illness in both parents. In those days, when no effective medicines existed for any form of psychiatric illness, most families bearing mental disease not surprisingly kept this knowledge as secret as possible. There must have been many couples, over-worried about producing mentally disturbed offspring, who chose not to have children.

Also of obsessive concern to the eugenicists were individuals with feeble-mindedness, where Davenport believed that genes were also involved. With his certainty that all children of feeble-minded parents would be defective, he wrote of the "folly, yes the crime, of letting two such persons marry."[2] In his mind, the inhabitants of rural poorhouses were there largely because of their feeble-mindedness, and he considered one of our nation's worst dangers to be the constant generation of feeble-minded individuals by the unrestrained lusts of parents of similar conditions.

It was to stop such further contaminations of the American germ plasm that Davenport, as early as 1911, saw the need for state control of the propagation of the mentally unstable or defective. Initially, he did not favor adoption of state laws allowing for their compulsory sterilization, an idea then considered wise and humane by much of that era's socially progressive elite. He feared that with pregnancy no longer a worry, the sexual urges of the sterilized, unstable mentally impaired might cause more harm to society than even the procreation of more of their kind. He wanted mentally impaired women to be effectively segregated (imprisoned?) from the impaired of the opposite sex until after they passed the age of procreation. This prescription, however, was totally unrealistic, and the American eugenics movement as a whole later promoted the compulsory sterilization legislation that spread to 30 states by the start of World War II.

If the eugenics movement instead had focused predominately on physically disabled as opposed to mentally disabled victims of "bad" genes, we might now be able to look back at it as a mixture of sloppy science and well-intentioned but kooky naiveté. Photos of the Eugenics Booths of 1920s State Farm Fairs are virtually laughable. In them can be seen "fitter families" displayed near the pens at which prize cattle were shown. The thought that sights of their earnest faces would lead to preferential procreation of more of the same now stretches our credulity.

In contrast, the words and actions of Harry P. Laughlin, Davenport's close associate and superintendent of the Eugenics Record Office, today can only make our minds flinch. Pleased that his ancestors were traceable to the American Revolution, Laughlin shared Davenport's belief that the strengths and weaknesses of national and religious groups were rooted in genetic as well as in cultural origins. While, at least in public, Davenport wrote that no individual should be refused admission to the United States on the basis of religious group or national origin, Laughlin stated as scientific fact before appropriate congressional bodies that the new Americans from Eastern and Southern Europe were marked by unacceptable amounts of insanity, mental deficiency, and criminality.[3] Though lacking any solid supporting evidence, he promoted the belief that the newest immigrants to our shores were much more likely to be found in prisons and insane asylums than were the descendants of earlier waves of English, Irish, German, and Scandinavian settlers. Even though the then current postwar hysteria against unrestrained immigration by itself might have led to the 1924 legislation, there is no doubt that Laughlin's testimony tilted the composition of the future immigrants to Northern Europeans.

Using the First IQ Tests to Justify Racial Discrimination within the United States

The emergence of intelligence measuring reinforced the belief of America's prosperous people that their wealth reflected their family's innate intellectual superiority. The French psychologist Alfred Binet (1857–1911) was the first person to try to systematically measure intelligence, responding to a 1904 request from the French government to detect mentally deficient children. The resulting Binet-Simon tests crossed the Atlantic by 1908, being first deployed in the United States by Henry Goddard in New Jersey at a training school for feeble-minded boys and girls. Soon afterward, he went on to test 2,000 children with a broad range of mental abilities. Initially, there was considerable public opposition to the testing of "normal" individuals because of the test's first use on the feeble-minded. Within only a few years, however, revised Binet-Simon tests, more appropriate for precocious children, were prepared by Lewis Ternan (1877–1956) at Stanford University. These so-called Intelligence Quotient (IQ) tests were soon employed on hundreds of thousands of army draftees during World War I. Their main function was not to weed out mental defectives, but to assign recruits to appropriate army roles.

Those administering the tests, led by the noted psychologist Robert M. Yerkes (1876–1956), claimed they were seeing native intelligence independent of the recruit's environmental history. Yet, clearly, many of the questions or arithmetic problems would be more easily answered by those with extensive schooling and possessing a broad vocabulary. Not surprisingly, the non-English-speaking recruits just off the immigration boats tested badly. Such "objective test data" further convinced the eugenicist world that not only was mental deficiency genetically determined, but so was general intelligence.

Although black men from urban areas tested higher than white Southern rural men, their IQ scores were significantly lower than their white equivalents from the same communities. Given that intelligence measurements virtually by necessity have cultural biases, the comparative data assembled from the army recruits had little real meaning. In many ways, it was like comparing oranges with apples. Nonetheless, the data summarized in "Psychological Examining in the United States Army" were used to justify the discriminatory segregation laws that effectively made America's black population second-class citizens. Genetic inequalities across so-called race boundaries were taken for granted, and 29 states maintained laws against black-white intermarriages, often using the argument that the superior white germ stock would be diluted with inferior genes.

Although eugenics had its origin in England, it never affected the national consciousness there as it did in the United States. With social class stratification so long a characteristic feature of British life, the ruling classes had no need of further justification for their privileged existence. Enthusiastic prewar eugenics movements, nonetheless, sprang up all over the continent, extending even to South America and Japan in the 1920s. Only in two European countries, Germany (under the National Socialists) and Sweden, was legislation enacted for obligatory sterilization of individuals thought to be the bearers of disabling genes.

Even before World War II started, eugenics in the United States was beginning to be perceived more as a social than a scientific movement. In 1930 the leaders of the Carnegie Institution of Washington had been told by leading American biologists that its Cold Spring Harbor Eugenics Record Station practiced sloppy, if not dishonest, science. But with its founder Charles Davenport nearing retirement, it was allowed to expire more slowly than in retrospect it should have. Its doors closed only when Milislav Demerec became director of the Department of Genetics in 1942.

There thus was the embarrassment of Harry Laughlin's receipt in 1936 of an honorary degree from the University of Heidelberg in recognition of his contributions to racial hygiene. Undoubtedly pleased that eugenics, then fading in the United States, was becoming even more ascendant in Germany, Laughlin went to New York to receive his diploma from the German Diplomatic Counsel. Only after the liberation first of Poland and then of Germany did the full horror of the racially based genocide policies of National Socialism finally become known, generating even further disgust for the pseudoscientific theories of race superiority and purity that underpinned Nazi ideology. Anyone subsequently calling himself a eugenicist put his reputation as a decent moral human at risk.

Eugenics, a Dirty Word, as the Search for the Chemical Nature of the Gene Begins

By the time I first came to Cold Spring Harbor for the summer of 1948, accompanying my Ph.D. supervisor, Salvador Luria, then a professor at Indiana University, the Eugenics Record Office had been virtually expunged from its consciousness. Only in the library was its ugly past revealed through the German journals of the 1920s and 1930s on Human Genetics and Race Hygiene. No one that summer showed any interest in human genetics as a science or toward the general question of how much of human behavior reflects nature as opposed to nurture. Instead, genetic research there focused on the fundamental nature of genes and their functioning. It was not that human genetic diseases had suddenly become unimportant to its director, Milislav Demerec. But there was general agreement, both by the year-round staff and the many summer visitors, that until the chemical identity of the gene was elucidated and the general pathways by which it controlled cell structure and functioning were known, it was premature to even speculate how genes contributed to human development and behavior.

Then, only five years were to pass before the gene was revealed in 1953 to be the double helix. By 1966 the genetic code was established, and soon after gene expression was seen to be controlled by DNA-binding regulatory proteins. Genetics, happily then, had no reasons to intersect politics, except in Russia, where the absurdity of its Lamarckian philosophy became painfully clearer to its intelligentsia with every new major advance in molecular genetics. These major genetic breakthroughs were largely accomplished using the simple genetic systems provided by bacteria and their viruses that go under the name "phages". Phage had become so well understood genetically by 1969 that it became possible to create specific phage strains cleverly engineered to carry specific bacterial genes from one bacterial strain to another. Although the phage transductional system proved not to be a forerunner for eventual human genetic engineering, this was not true for the much more powerful and general "recombinant DNA" methodologies that Herbert Boyer and Stanley Cohen developed four years later, in 1973, just 20 years after the discovery of the double helix. Their new procedures allowed the isolation (cloning) of specific genes, through their insertion into tiny chromosomes (plasmids) that could be moved from one cell to another.

RECOMBINANT DNA MOLECULES AS POTENTIAL BIOHAZARDS

The resulting recombinant DNA era, however, despite all the promises it held for major scientific advances, did not immediately take off. It initially stalled because of fears among some scientists that many new forms of DNA created in the laboratory could include some that would pose unacceptable dangers to life as it now exists. In particular was the fear that highly pathogenic new forms of viruses and bacteria might be created. To give time to assess such potential dangers scientifically, a scientist-initiated moratorium on recombinant DNA research was declared in 1975.

During the moratorium, which was strictly held to over the next two years, governmental committees were set up in the United States and in various European countries to assess the potential dangers from recombinant DNA experimentation in relation to its potential benefits for biology, medicine, and agriculture. No plausible scientific reasons for stopping such research emerged, and such committees, often containing public as well as scientific representatives, invariably concluded that in the absence of any quantifiable potential dangers, it would be irresponsible not to move ahead with experiments that could dramatically change the nature of biology.

Although there was no evidence of danger, there nonetheless arose much visible and sometimes regretfully effective opposition to recombinant DNA research. Here the distinction should be made between objections from scientists who understand the technical issues involved and opposition from groups of public citizens who, though not understanding the science involved, nonetheless oppose much to all recombinant DNA research. Although some initial opposition arose from scientists, many of whose own DNA research was not going well, virtually all the continuing scientific opponents in their heart had left-wing political hang-ups. As leftists, they did not want genes involved in human behavioral differences and feared that the onslaught of scientific advances that would follow from the unleashing of recombinant DNA might eventually allow genes affecting mental performance to be isolated and studied.

Political ideologies by themselves, however, were not the main cause of the prolonged and sometimes very effective opposition to recombinant DNA in Germany. Even today, much of Germany's reluctance to accept genetic engineering lies in the postwar fate of Hitler's biological conspirators. Naively, as outsiders we long assumed that they would have all been treated as potential if not real war criminals, with even those of only slight guilt, losing further opportunities for academic leadership if not existence. But as the German geneticist Benno Müller-Hill courageously pointed out in his 1984 book, *Todlicle Wissenschaff* (Murderous Science), there was no attempt by the German academic community to find out what truly happened.[4] Instead, it became academically dangerous in Germany to explore the half-truths that allowed many key practitioners of Nazi eugenics to resume important academic posts. Though a number of professors who early on joined the Nazi Party or SS and were directly involved with its genocide programs committed suicide, there were many Nazi-assisting scientists who slid quietly back into academic prominence, successfully claiming that they were only apolitical advisors.

The most damning example was that of Professor Otmar von Verschuer, who actively helped the Nazis, beginning at the Kaiser Wilhelm Institute for Anthropology, Human Genetics, and Eugenics, which was built with funds from the Rockefeller Foundation in 1926.[5] Earlier involved in distinguishing Jews and part-Jews, von Verschuer later closely collaborated with his former assistant, the now-notorious Joseph Mengele, then doing "scientific" research at Auschwitz. After the war, he nonetheless was appointed to be professor of human genetics at the University of Munster. Equally disturbing was the postwar appointment of Fritz Lenz as head of an institute for the study of human heredity at the University of Gottingen, Germany's most distinguished university. Although clearly a very competent scientist, he was a major advisor for laws on euthanasia between 1939 and 1941, as well as author of a 1940 memorandum, "Remarks on resettlement from the point of view of guarding the race."[6]

The postwar 1949 exoneration of von Verschuer occurred despite knowledge of the 1946 article in *Die New Zeit*, accusing him of studying eyes and blood samples sent to him from Auschwitz by Joseph Mengele. Yet a committee of professors, including Professor Adolf Butenandt, later the head of the Max Planck Gesellschaft (the postwar name for the Kaiser Wilhelm Gesellschaft), concluded that von Verschuer, who possessed all the qualities appropriate for a scientific researcher and teacher of academic youth, should not be judged on a few isolated events of the past. I find it difficult to believe that the Butenandt committee had gone to the trouble of reading von Verschuer's article published in the *Volkischer Beobachter* in 1942. In it he wrote, "Never before in the course of history has the political significance of the Jewish question emerged so clearly as it does today. Its definitive solution as a global problem will be determined during the course of this war."[7] There may be more reason to remember Professor Butenandt for his part in the von Verschuer whitewash than for his prewar Nobel Prize for research on the chemistry of the estrogen sex hormone.

KEEPING GOVERNMENTS OUT OF GENETIC DECISIONS

No rational person today should have doubts whether genetic knowledge properly used has the capacity to improve the human condition. For example, knowing what is wrong at the molecular level during psychosis should greatly enhance our capacity to develop drugs that will stop schizophrenia and bipolar disease. Other genetic disabilities may, with time, be neutralized by so-called gene therapy procedures, restoring normal cell functioning by adding good copies of the missing normal genes.

Although gene therapy enthusiasts have promised too much for the near future, it is difficult to imagine that they will not, with time, cure some unwanted genetic conditions.

For the time being, however, we may have to place most of our immediate hopes for reducing the human burden of genetic diseases on the use of antenatal diagnostic procedures. They increasingly will let us know whether a fetus is carrying a mutant gene that will seriously proscribe its eventual development into a functional human being. But even among individuals who firmly place themselves on the pro-choice side and do not want to limit women's rights for abortion, opinions frequently are voiced that decisions obviously good for individual persons or families may not always be appropriate for the societies in which we live. For example, by not wanting to have a physically or mentally handicapped child or one who would have to fight all its life against possible death from cystic fibrosis, are we not reinforcing the second-rate status of such handicapped individuals? And what would be the consequences of isolating genes that give rise to the various forms of dyslexia, opening up the possibility that women will take antenatal tests to see if their prospective child is likely to have a bad reading disorder? Is it not conceivable that such tests would lead to our devoting less resources to the currently reading-handicapped children whom now we accept as an inevitable feature of human life? Here I'm afraid that the word handicap cannot escape its true definition—being placed at a disadvantage. From this perspective, seeing the bright side of being handicapped is like praising the virtues of extreme poverty.

Only harm, I fear, will come from any form of society-based restriction on individual genetic decisions. Decisions from committees of well-intentioned individuals will all too often emerge as vehicles for seeming to do good as opposed to doing good. Moreover, we should necessarily worry that once we let governments tell their citizens what they cannot do genetically, we must fear they also have power to tell us what we must do. But for us as individuals to feel comfortable making decisions that affect the genetic makeup of our children, we correspondingly have to become genetically literate.

ETHICS WITHIN THE HUMAN GENOME PROJECT

The moment I began in October 1988 my almost four-year period of helping lead the Human Genome Project in its primary mission of sequencing the 3×10^9 base pair human genome, I stated that 3 percent of the funds provided by the National Institutes of Health should support research and discussion on the Ethical, Legal, and Social Implications (ELSI) of the new resulting genetic knowledge. A lower percentage might be seen as tokenism, while I then could not see wise use of a larger sum. Under my 3 percent proposal (later raised by Congress to 5 percent), some $6 million (3 percent of $200 million) would eventually be so available, a much larger sum than ever before provided by our government for the ethical implications of biological research. In putting ethics so soon into the genome agenda, I was responding not only to my long-term interest in how genetics is used for the moral benefit of human beings but also to my own fear that, all too soon, critics of the Human Genome Project would point out that I was a representative of the Cold Spring Harbor

Laboratory that once housed the controversial Eugenics Record Office. My not forming a genome ethics program quickly might be falsely used as evidence that I was a closet eugenicist, having as my real purpose the finding of human genes that would justify class and race-based discriminatory practices.

Complicating the subsequent ELSI agendas has been the increasingly obvious fact that different genetic diseases present quite different spectra of ethical, legal, and social dilemmas. And nobody has specifically come up with universally acceptable regulations or laws protecting the privacy of individual genetic records. Even when appropriate satisfactory laws and regulations are in place, there will still be many dilemmas that cannot easily be handled by these means. What responsibility, for example, do individuals have to learn about their genetic makeup prior to their parenting of children? In the future, will we be generally regarded as morally neglectful when we knowingly permit the birth of children with severe genetic defects? And do such victims later have legal recourse against their parents who have taken no action to help prevent their coming into the world with few opportunities of living a life without pain and emotional suffering?

THE MISUSE OF GENETICS BY HITLER MUST NOT RESTRICT ITS HORIZONS IN THE FUTURE

Because of Hitler's use of the term "Master Race", we should not feel the need to say that we never want to use genetics to make humans more capable than they are today. The idea that genetics could or should be used to give humans power that they do not now possess, however, strongly upsets many individuals first exposed to the notion. I suspect such fears, in some ways, are similar to concerns now expressed about the genetically handicapped of today. If more intelligent human beings might someday be created, would we not think less well about ourselves as we exist today? Yet anyone who proclaims that we are now perfect as humans has to be a silly crank. If we could honestly promise young couples that we knew how to give them offspring with superior character, why should we assume they would decline? Those at the top of today's societies might not see the need. But if your life is going nowhere, shouldn't you seize the chance of jump-starting your children's future?

Those of us who venture forth into the public arena to explain what genetics can or cannot do for society seemingly inevitably come up against individuals who feel that we are somehow the modern equivalents of Hitler. Here we must not fall into the absurd trap of being against everything Hitler was for. It was in no way evil for Hitler to regard mental disease as a scourge on society. Almost everyone then, as is still true today, was made uncomfortable by psychotic individuals. It is how Hitler treated German mental patients that still outrages civilized societies and lets us call him immoral. Genetics per se can never be evil. It is only when we use or misuse it that morality comes in. That we want to find ways to lessen the impact of mental illness is inherently good. The killing by the Nazis of the German mental patients for reasons of supposed genetic inferiority, however, was barbarianism at its worst.

The concept of genetic determinism is inherently unsettling to the human psyche, which likes to believe that it has some control over its fate. No one feels comfortable with the thought that we, as humans, virtually all contain one to several "bad" genes that are likely to limit our abilities to fully enjoy our lives. Nor do we necessarily take pleasure in the prospect that we will someday have gene therapy procedures that will let scientists enrich the genetic makeup of our descendants. Instead, there has to be genuine concern as to whether our children or their governments decide what genes are good for them.

Genetics as a discipline must thus strive to be the servant of the people, as opposed to governments, working to mitigate the genetic inequalities arising from the random mutations that generate our genetic diseases. Never again must geneticists be seen as the servants of political and social masters who need demonstrations of purported genetic inequality to justify their discriminatory social policies.

NOTES

Portions of this chapter were previously published in *A Passion for DNA: Genes, Genomes and Society*, pp. 3–5, 179–208, 209–222, by James D. Watson, Cold Spring Harbor Laboratory Press, 2000.

1. Charles Benedict Davenport, *Heredity in Relation to Eugenics* (New York: Henry Holt and Company, 1911).
2. Ibid., 67.
3. Cold Spring Harbor Laboratory, "Immigration," Dolan Learning Center Image Archive on the American Eugenics Movement, http://www.eugenicsarchive.org/html/eugenics/static/themes/10.html.
4. Benno Müller-Hill, *Murderous Science: Elimination by Scientific Selection of Jews, Gypsies, and Others in Germany, 1933–1945* (Cold Spring Harbor, New York: Cold Spring Harbor Laboratory Press, 1997). My interview and tour of the gas chamber at Bernberg Psychiatric Hospital with Benno Müller-Hill can be seen at http://www.dnalc.org/mediashowcase/index.html?q=tour&page=4.
5. Paul Weindling, *Health, Race and German Politics between National Unification and Nazism* (New York: Cambridge University Press, 1989), 432–439.
6. Müller-Hill, *Murderous Science*, 83.
7. Ibid., 220.

CHAPTER 8

THE STAIN OF SILENCE: NAZI ETHICS AND BIOETHICS

ARTHUR L. CAPLAN

THE ROLE OF THE HOLOCAUST IN THE HISTORY OF BIOETHICS

There is little agreement among scholars of the history of bioethics about where bioethics begins. It is understood that medical ethics long predates bioethics, but medical ethics, whether of the early Greek physicians or of the nineteenth-century physicians and surgeons of Britain and the United States, consists of a code or professional conduct for doctors. Bioethics is far more than that. It is the study of ethical problems in the practice of medicine involving doctors, nurses, other health care professionals, managers, payers, patients, and researchers. Bioethics' scope ranges beyond clinical medicine to include moral problems in the life sciences, research, and health economics. More to the point, bioethics involves multidisciplinary reflections on ethical problems, not simply the wisdom of the medical profession about how best to practice.

Most scholars date the rise of bioethics to the early 1970s. At that time, challenges arose for medicine in America that involved an increasing unease with physician paternalism, uncertainty over how to manage emerging technologies such as ventilators and dialysis, disputes over the morality of genetic testing and elective abortion, and controversy about scandals being discovered in American medical research.[1]

But some place the origins of bioethics much earlier. They see bioethics arising out of the ashes of the Holocaust.[2] And indeed, the Nuremberg Code is frequently held up in courses and textbooks on medical ethics as the quintessential bioethical document—a kind of "constitution" of bioethics. But this document almost always appears in books, anthologies, and courses out of context. Very little is said about the actual experiments in concentration camps that generated this document. And even less is said about the moral rationales those involved in the horrific research gave in their defense. Why?

If bioethics truly arose as a result of the Holocaust, then one would expect to encounter much more discussion and debate over what happened in the camps than is, in fact, the case. Indeed, it was not until the early 1990s that any formal meeting or conference was held on the Holocaust and bioethics by any program, institute, or center involved in bioethics. Nor are there many citations, even to this day, in discussions of bioethics to anything other than the Nuremberg Code.

It would seem that if bioethics really had its roots in the Holocaust, they did not run deep. Nor have they shaped the way the field has approached the analysis and resolution of ethical problems in any serious way. Why is this so?

Why the Loud Silence about the Holocaust?

The events of the Holocaust are so horrid that they speak for themselves. What more is there to say about mass murder and barbaric experimentation except that it was unethical? So it may be that some bioethicists have had nothing to say about these events simply because they may believe there is nothing that can be said about them. But such an excuse cannot be applied to the research done in the concentration camps. There, medical people did horrid things to human beings but did so in the name of gaining new knowledge. How could these brutal and lethal experiments elicit little comment from bioethicists?

Many scholars of the camp experiments have dismissed the research conducted there as worthless. Those involved have been dismissed as political hacks, lunatics, and crackpots. What point is there in discussing the ethics of research that is nothing more than torture disguised as science?[3]

Yet another reason for the failure to grapple with Nazi camp experiments is the tradition of trying to offer psychological explanations for the behavior of those involved in the killing. A psychological account may make moral explanations seem unnecessary.

How could doctors kill and maim with abandon? Many psychiatrists argue that those who went to work at the gas chambers and dissection rooms did so through adaptations of personality and character that make their conduct understandable. They split their personalities and basically lived an almost dual personality existence. This may be true, but such accounts make it difficult to hold those doctors involved in the camps morally accountable for their conduct.[4]

And there is always the fear that to talk of the ethics of the research done in the camps is to lend barbarism a convenient disguise. It is simply wrong to look at the ethical justifications for what was done because it confers a false acceptability on what was manifestly wrong.

Perhaps the most important reason for the absence of commentary on the ethics of the research done in the camps is that such questions open a door that few bioethicists wish to enter. If moral justifications can be given for why someone deemed mass murder appropriate in the name of public health or thought that it was right to freeze hapless men and women to death or decompress them or infect them with lethal doses of typhus, then to put the question plainly, what good is bioethics?

DEBUNKING THE MYTHS OF INCOMPETENCE, MADNESS, AND COERCION

It is comforting to believe that health care professionals from the nation that was the world's leader in medicine at the time, who had pledged an oath to "do no harm," could not conduct brutal, often lethal, experiments upon innocent persons in concentration camps. It is comforting to think that it is not possible to defend wound research on the living in moral terms. It is comforting to think that anyone who espouses racist, eugenic ideas cannot be a competent, introspective physician or scientist. Nazi medical crimes show that each of these beliefs is false.[5]

It is often believed that only madmen, charlatans, and incompetents among doctors, scientists, public health officials, and nurses could possibly have associated with those who ran the Nazi party. Among those who did their "research" in Auschwitz, Dachau, and other camps, some had obvious psychological problems, were lesser scientific lights, or both.[6] But, there were also well-trained, reputable, and competent physicians and scientists who were also ardent Nazis. Some conducted experiments in the camps. Human experimentation in the camps was not conducted only by those who were mentally unstable or on the periphery of science. Not all who engaged in experimentation or murder were inept or ill trained.[7]

Placing all of the physicians, health professionals, and scientists who took part in the crimes of the Holocaust on the periphery of medicine and science allows another myth to flourish—that medicine and science went "mad" when Hitler took control of Germany. Competent and internationally renowned physicians and scientists could not willingly have had anything to do with Nazism. However, the actions as well as the beliefs of German physicians and scientists under Nazism stand in glaring contrast to this myth.[8]

Once identified, the myths of incompetency and madness make absolutely no sense. How could flakes, crackpots, and incompetents have been the only ones supporting Nazism? Could the Nazis have had any chance of carrying out genocide on a staggering, monumental scale against victims scattered over half the globe without the zealous help of competent biomedical and scientific authorities? The technical and logistical problems of collecting, transporting, exploiting, murdering, scavenging, and disposing of the bodies of millions from dozens of nations required competence and skill, not ineptitude and madness.

The Holocaust differs from other instances of genocide in that it involved the active participation of medicine and science. The Nazis turned to biomedicine specifically for help in carrying out genocide after their early experience using specially trained troops to murder in Poland and the Soviet Union proved impractical.[9]

Another myth that has flourished in the absence of a serious analysis of the moral rationales proffered by those in German biomedicine who participated in the Holocaust is that those who participated were coerced. Many doctors, nurses, and scientists in Germany and other nations have consoled themselves about the complicity of German medicine and science in genocide with the fable that, once the Nazi regime seized power, the cooperation of the biomedical and scientific establishments was only secured by force.[10] Even then, this myth has it, cooperation among doctors, scientists, and public health officials with Nazism was grudging.

The myths of incompetence, madness, and coercion have obscured the truth about the behavior of biomedicine under Nazism. Most of those who participated did so because they believed it was the right thing to do. This helps to explain the relative silence in the field of bioethics about both the conduct and justifications of those in biomedicine who were so intimately involved with the Nazi state.[11]

Why Does Bioethics Have So Little to Say about the Holocaust?

If one dates the field of bioethics from the creation of the first bioethics institutes and university programs in the United States in the mid-1970s, then the field is roughly 35 years old. Incredibly, no book-length bioethical study exists examining the actions, policies, abuses, crimes, or rationales of German doctors and biomedical scientists.

There has been almost no discussion of the roles played by medicine and science during the Nazi era in the bioethics literature. Rather than see Nazi biomedicine as morally bad, the field of bioethics has either remained silent in the face of these crimes or accepted the myths that Nazi biomedicine was inept, mad, or coerced.

By subscribing to these myths, bioethics has been able to avoid a painful confrontation with the fact that many of those who committed the crimes of the Holocaust were competent physicians and scientists who acted from strong moral convictions. When called to account at Nuremburg and other trials for their actions, Nazi doctors, scientists, and public health officials were surprisingly forthright about their reasons for their conduct.

The puzzle of how it came to be that physicians and scientists who committed so many crimes and caused so much suffering and death did so in the belief that they were morally right cries out for analysis, discussion, and debate. But it is tremendously painful for those in bioethics to have to undertake such an analysis.

It if often presumed, if only tacitly, by those who teach bioethics that those who know what is ethical will not behave in immoral ways. What is the point of doing bioethics, of teaching courses on ethics to medical, nursing, and public health students if the vilest and most horrendous of deeds and policies can be justified by moral reasons? Bioethics has been speechless in the face of the crimes of Nazi doctors and biomedical scientists precisely because so many of these doctors and scientists believed they were doing what was morally right to do.

Experimentation In the Camps

At least 26 different types of experiments were conducted for the explicit purpose of research in concentration camps or using concentration camp inmates in Germany, Poland, and France during the Nazi era.[12] Among the studies in which human beings were used in research were the analysis of high-altitude decompression on the human body; attempts to make sea water drinkable; the efficacy of sulfanilamide for treating gunshot wounds; the feasibility of bone, muscle, and joint transplants; the ability to treat burns caused by incendiary bombs; the efficacy of polygal for treating trauma-related bleeding; the efficacy of high-dose radiation in causing sterility; the efficacy of phenol (gasoline) injections as a euthanasia agent; the efficacy of electroshock therapy; the symptoms and course of noma (starvation caused skin gangrene); the postmortem examination of skeletons and brains to assess the effects of starvation; the efficacy of surgical techniques for sterilizing women; and the impact of stress and starvation on ovulation, menstruation, and cancerous growths in the reproductive organs of women. A variety of other studies were carried out on twins, dwarves, and those with congenital defects. Some camp inmates were used as subjects to train medical students in surgery.

The question of whether any of these activities carried out in the name of medical or scientific research upon unconsenting, coerced human beings deserves the label of "research" or "experimentation" is controversial.[13] When the description of research is broadened further to include the intentional killing of human beings in order to establish what methods are most efficient, references to "research" and "experimentation" begin to seem completely strained. Injecting a half-starved young girl with phenol to see how quickly she will die or trying out various forms of phosgene gas on camp inmates in the hope of finding cheap, clean, and efficient modes of killing so the state can effectively prosecute genocide is not the sort of activity associated with the term "research."

But murder and genocide are not the same as intentionally causing someone to suffer and die to fulfill a scientific goal. Killing for scientific purposes, while certainly as evil as murder in the service of racial hygiene, is, nonetheless, morally different. The torture and killing that were at the core of Nazi medical experiments involve not only torture and murder but also the exploitation of human beings to serve the goals of science. To describe what happened in language other than that of human experimentation blurs the nature of the wrongdoing. The evil inherent in Nazi medical experimentation was not simply that people suffered and died but that they were exploited for science and medicine as they died.

A summary report prepared for the American military about the hypothermia experiments has been cited in the peer-reviewed literature of medicine more than two dozen times since the end of World War II. Not only was the data examined and referenced, it was applied. British air-sea rescue experts used the Nazi data to modify rescue techniques for those exposed to cold water.[14] The force of the question, "should the data be used?" is diminished not only because there are reasons to doubt the reliability and exclusivity of the data, but because the question has already been answered—Nazi data has been used by many scientists from many nations.

The German researchers involved in the camp studies believed in their work. A few protested attempts by the prosecution, in its effort to highlight the barbarity of what they had done, to demean, or to disparage the caliber of their research. No one apologized for their role in various experiments conducted in the camps. Instead, those put on trial attempted to explain and justify what they had done, often couching their defense in explicitly moral terms.

THE ETHICS OF EVIL

The first group of individuals to be put on trial by the Allies at the end of the war were physicians and public health officials. The role they had played in conducting or tolerating cruel and often lethal experiments in the camps dominated the trials.[15]

A review of the major moral arguments presented by defendants at the Nuremberg trials sheds light not only on the moral rationales that were given for the hypothermia and phosgene gas experiments but, also, for the involvement of biomedicine in the broad sweep of what the prosecution termed "crimes against humanity."[16]

One of the most common moral rationales given at the trials was that no wrong had been done because those who were subjects had volunteered. Prisoners might be freed, some defendants argued, if they survived the experiments. The prospects of release and pardon were mentioned very frequently during the trial since they were the basis for the claim that people participated voluntarily in the experiments (Nuremberg Trial Transcripts, 1946–1947). On this line of thinking, experimentation was justified because it might benefit the subjects. The major flaw with this moral rationale is simply that it was false—no subject was ever freed.

Another of the key rationales on the part of those put on trial was that only people who were doomed to die were used for biomedical purposes (Nuremberg Trial Transcripts, 1946–1947). Time and again the doctors who froze screaming subjects to death or watched their brains explode as result of rapid decompression stated that only prisoners condemned to death were used. It seemed morally defensible to physicians and scientists to learn from what they saw as the inevitable deaths of camp inmates.

A third ethical rationale for performing brutal experiments upon innocent subjects was that participation in lethal research offered expiation to the subjects. By being injected, frozen, or transplanted, subjects could cleanse themselves of their crimes. Suffering prior to death as a way to atone for sin seemed to be a morally acceptable rationale for causing suffering to those who were guilty of crimes.

The problem with this ethical defense is that those who were experimented on or made to suffer by German physicians and scientists were never guilty of any crime other than that of belonging to a despised ethnic or racial minority or for holding unacceptable political or religious views.

A fourth moral rationale, one that is especially astounding even by the standards of self-delusion in evidence throughout the trial proceedings, was that scientists and physicians had to act in a value-neutral manner. They maintained that scientists and doctors are not responsible for and have no expertise about values and, thus, could not be held accountable for their actions.

> [I]f the experiment is ordered by the state, this moral responsibility of experimenter toward the experimental subject relates to the way in which the experiment is performed, not the experiment itself.
>
> (Nuremberg Trial Transcripts, 1946–1947)

Some researchers only felt themselves responsible for the proper design and conduct of their research. They felt no moral responsibility for what had occurred in the camps because they did not have any expertise concerning moral matters. They claimed to have left decisions about these matters to others. In other words, they argued that scientists in order to be scientists could not take normative positions about their science.

The fifth moral justification for what had happened presented by many of the defendants was that they had done what they did for the defense and security of their country. All actions were done to preserve the Reich during "total" war (Nuremberg Trial Transcripts, 1946–1947).

> Germany was engaged in war at that time. Millions of soldiers had to give up their lives because they were called upon to fight by the state. The state employed the civilian population for work according to state requirements. The state ordered employment in chemical factories which was detrimental to health.... In the same way the state ordered the medical men to make experiments with new weapons against dangerous diseases.
>
> (Nuremberg Trial Transcripts, 1946–1947)

Total war, war in which the survival of the nation hangs in the balance, justifies exceptions to ordinary morality, the defendants maintained. Allied prosecutors had much to ponder in thinking about this defense in light of the fire-bombing of Dresden, Tokyo, and the dropping of nuclear bombs on Hiroshima and Nagasaki.

The last rationale is the one that appears to carry the most weight among all the moral defenses offered. Many who conducted lethal experiments argued that it was reasonable to sacrifice the interests of the few to benefit the majority.

The most distinguished of the scientists who was put on trial, Gerhard Rose, the head of the Koch Institute of Tropical Medicine in Berlin, said that he initially opposed performing potentially lethal experiments to create a vaccine for typhus on camp inmates. But he came to believe that it made no sense not to risk the lives of 100 or 200 men in pursuit of a vaccine when 1,000 men a day were dying of typhus on the Eastern front. What, he asked, were the deaths of 100 men compared to the possible benefit of getting a prophylactic vaccine capable of saving tens of thousands? Rose, because he admitted that he had anguished about his own moral duty when asked by the *Wehrmacht* to perform the typhus experiments in a concentration camp, raises the most difficult and most plausible moral argument in defense of lethal experimentation.

Justifying the sacrifice of the few to benefit the majority is a position that must be taken seriously as a moral argument. In the context of the Nazi regime, it is fair to point out that sacrifice was not borne equally by all as is true of a compulsory draft that allows no exceptions. It is also true that many would argue that no degree of benefit should permit intrusions into certain fundamental rights.

Crude utilitarianism is a position that sometimes rears its head in contemporary bioethical debate. For example, some argue that we ought not spend scarce social resources on certain groups within our society such as the elderly so that other groups, such as children, may have greater benefits. Those who want to invoke the Nazi analogy may be able to show that this form of crude utilitarian thinking does motivate some of the policies or actions taken by contemporary biomedical scientists and health care professionals, but they need to do so with great caution.

In closely reviewing the statements that accompany the six major moral rationales for murder, torture, and mutilation conducted in the camps—freedom was a possible benefit, only the condemned were used, expiation was a possible benefit, a lack of moral expertise, the need to preserve the state in conditions of total war, and the morality of sacrificing a few to benefit many—it becomes clear that the conduct of those who worked in the concentration camps was at least to some extent guided by moral rationales. It is also clear that all of these moral arguments were nested within a biomedical interpretation of the danger facing Germany and a framework of vicious racism.

Physicians could justify their actions, whether direct involvement with euthanasia and lethal experiments or merely support for Hitler and the Reich, on the grounds that the Jew, the homosexual, the congenitally handicapped, and the Slav posed a threat, a biological threat, a genetic threat, to the existence and future of the Reich. The appropriate response to such a threat was to eliminate it, just as a physician must eliminate a burst appendix by means of surgery or dangerous bacteria by using penicillin.[17]

Viewing specific ethnic groups and populations as threatening the health of the German state permitted, and in the view of those on trial demanded, the involvement of medicine in mass genocide, sterilization, and lethal experimentation. The biomedical paradigm provided the theoretical basis for allowing those sworn to the Hippocratic principle of non-maleficence to kill in the name of the state.

The Neglect of the Holocaust and Nazism in Bioethics

Why has the field of bioethics not attended more closely to the Holocaust and the role played by German medicine and science in the Holocaust? Why have the moral arguments bluntly presented by the Nazis received so little attention? These are questions that do not admit of simple answers.

The crimes of doctors and biomedical scientists revealed at the Nuremberg trials were overwhelming in their cruelty. Physicians and scientists supervised and, in some cases, actively participated in the genocide of millions, directly engaged in the torture of thousands, and provided the scientific underpinning for genocide. Hundreds of thousands of psychiatric patients and senile elderly persons were killed under the direct supervision of physicians and nurses. Numerous scientists and physicians, some of whom headed internationally renowned research centers and hospitals, engaged in cruel and sometimes lethal experiments on unconsenting inmates of concentration camps.

Ironically, the scale of immorality is one of the reasons why the moral reasoning of health care professionals and biomedical scientists during the Nazi era has received little attention from contemporary bioethics scholars. It is clear that what Nazi doctors, biologists, and public health officials did was immoral. The indisputable occurrence of wrongdoing suggests that there is little for the ethicist to say.

Guilt by association has also played a role in making some bioethicists shy away from closely examining what medicine and science did during the Nazi era. Many doctors and scientists who were contemporaries of those put on trial at Nuremberg denied any connection between their own work or professional identities and those of the defendants in the dock. Contemporary doctors and scientists are, understandably, even quicker to deny any connection between what Nazi doctors or scientists did and their own activities or conduct. Many scholars and health care professionals, in condemning the crimes committed in the name of medicine and the biomedical sciences during the Holocaust, insist that all those who perpetrated those crimes were aberrant, deviant, atypical representatives of the health and scientific professions. Placing these acts and those who did them on the fringe of biomedicine keeps a needed distance between then and now.

Yet, by saying little and thereby allowing all Nazi scientists and doctors to be transformed into madmen or monsters, bioethicists ignore the fact that the Germany of the first half of this century was one of the most "civilized," technologically advanced, and scientifically sophisticated societies on the face of the globe. In medicine and biology, pre-World War II Germany could easily hold its own with any other scientifically literate society of that time. Indeed, the crimes carried out by doctors and scientists during the tenure of the Third Reich are all the more staggering in their impact and are all the more difficult to interpret precisely because Germany was such a technologically and scientifically advanced society.

The Holocaust is the exemplar of evil in our century. The medical crimes of that time stand as the clearest examples available of moral wrongdoing in biomedical science. Bioethics may have been silent precisely because there seems to be nothing to say about an unparalleled biomedical immorality, but silence leads to omission.

By saying little about the most horrific crimes ever carried out in the name of biomedicine and the moral views that permitted these crimes to be done, bioethics contributes to the most dangerous myth of all—that those engaged in evil cannot do so motivated by ethical beliefs. The challenge to bioethics and, indeed, all of ethics is to subject the beliefs that led to such horror to close critical scrutiny. It must be said that the ethical arguments advanced by the Nazis have some power. It should have been and remains the task of contemporary bioethics to show why that power is not persuasive and how bioethics has tried to develop norms and codes in direct response to the rationales the Nazis attempted to construct for the Holocaust.

NOTES

Portions of this chapter are adapted from my essay "The Ethics of Evil: The Challenge and the Lessons of Nazi Medical Experiments" in *Dark Medicine: Rationalizing Unethical Medical Research*, ed. William R. LaFleur et al. (Bloomington: Indiana University Press, 2007), 50–64.

1. Albert R. Jonsen , *A Short History of Medical Ethics* (New York: Oxford University Press, 2000).
2. George J. Annas, *American Bioethics: Crossing Human Rights and Health Law Boundaries* (New York: Oxford University Press, 2005).
3. Robert L. Berger, "Nazi Science: The Dachau Hypothermia Experiments," *New England Journal of Medicine* 322 (1990): 1435–1440.
4. Robert J. Lifton, *The Nazi Doctors: Medical Killing and the Psychology of Genocide* (New York: Basic Books, 1986); and Christopher Browning, *Ordinary Men: Reserve Police Battalion 101 and the Final Solution in Poland* (New York: Harper Collins, 1993).
5. Arthur L. Caplan, "Medicine's Shameful Past," *The Lancet* 363 (2004): 1741–1742; and Ulf Schmidt, *Karl Brandt: The Nazi Doctor: Medicine and Power in the Third Reich* (London: Hambledon Continuum, 2007).
6. Lifton, *The Nazi Doctors*.

7. Michael H. Kater, *Doctors Under Hitler* (Chapel Hill: The University of North Carolina Press, 1989); Robert N. Proctor, *Racial Hygiene: Medicine Under the Nazis* (Cambridge: Harvard University Press, 1992); Arthur L. Caplan, *When Medicine Went Mad: Bioethics and the Holocaust* (Totowa, NJ: Humana Press, 1992); Caplan, "Medicine's Shameful Past," 1741–1742; and Ulf Schmidt, *Karl Brandt.*
8. Proctor, *Racial Hygiene*; Kater, *Doctors Under Hitler*; and Schmidt, *Karl Brandt.*
9. Browning, *Ordinary Men.*
10. Lifton, *The Nazi Doctors*; Proctor, *Racial Hygiene*; and Kater, *Doctors Under Hitler.*
11. Arthur L. Caplan, Susan Benedict, and T. L. Page. Duty, "'Euthanasia': The Nurses of Meseritz-Obrawalde," *Nursing Ethics* 14, no. 6 (2007): 781–794.
12. Caplan, *When Medicine Went Mad.*
13. Berger, "Nazi Science," 1435–1440.
14. Caplan, *When Medicine Went Mad.*
15. Ibid.
16. Nuremberg Trial Transcripts, Case 1, The Medical Case (U.S.A. v. Karl Brandt et al., also known as The Doctors' Trial), 1946–1947, http://nuremberg.law.harvard.edu/php/docs_swi.php?DI=1&text=transcript, (hereafter cited in text).
17. Caplan, "Medicine's Shameful Past," 1741–1742.

CHAPTER 9

THE LEGACY OF THE NUREMBERG DOCTORS' TRIAL TO AMERICAN BIOETHICS AND HUMAN RIGHTS

GEORGE J. ANNAS

In this chapter, I will argue that modern bioethics was born at the Nuremberg Doctors' Trial, a health law trial that produced one of the first major human rights documents: the Nuremberg Code. Accepting this conclusion has significant consequences for contemporary American bioethics generally, and specifically in the context of the global war on terror in which the United States has used physicians to facilitate in interrogations, torture, and force-feeding hunger strikers.

The primary force shaping the agenda, development, and current state of American bioethics has not been either medicine or philosophy, but law, best described as health law. Like bioethics, health law is an applied field—in this case, law applied to medicine, biotechnology, and public health. Often the legal issues are raised in the context of a constitutional dispute, as in public debates over abortion, quarantine, the right to refuse treatment, and physician-assisted suicide; other times health law involves the more routine application of common law principles to new technologies or techniques, as in medical malpractice litigation; and still other times it is in the form of a debate over the wisdom or effectiveness of statutes and regulations, as in human experimentation, drug safety, patient safety, and medical practice standards.

American bioethics has had a major positive impact on the way medicine is currently practiced in the United States, especially in the areas of care of dying patients, including advance directives (living wills and health care proxies) and ethics committees as well as the establishment of rules governing medical research, including federal regulations to protect research subjects and institutional review boards (IRBs).

American bioethics has probably exhausted what it can usefully accomplish in these limited spheres. In the only other major area it has worked in, the related fields of abortion, embryo research, and cloning, it has had no real impact in debates that have been dominated by religion. Given this, I think it is fair to conclude that American bioethics is likely to have no real world future without a significant re-orientation of its focus and direction. I will suggest that the most useful reformulation involves recognition and engagement with two interrelated forces reshaping the world and simultaneously providing new frameworks for ethical analysis and action—globalization and public health. Most relevant for American bioethics is that globalization brings with it a new focus on international human rights law and its aspirations as articulated in the Universal Declaration of Human Rights.

Nuremberg and Bioethics

The boundaries between bioethics, health law, and human rights are permeable, and border crossings, including crossings by blind practitioners, are common. Two working hypotheses form the intellectual framework of this chapter: we can more effectively address the major health issues of our day if we harmonize all three disciplines, and American bioethics can be reborn as a global force by accepting its roots in the 1946–1947 Nuremberg Doctors' Trial and actively engaging in a health and human rights agenda. That these disciplines have often viewed each other with suspicion or simple ignorance tells us only about the past. They are most constructively viewed as integral, symbiotic parts of an organic whole.

Both American bioethics and international human rights were born of World War II, the Holocaust, and the Nuremberg tribunals. While the Doctors' Trial was only a part of Nuremberg and the new field of international human rights law, I believe it is accurate to conclude that the trial itself marked the birth of American bioethics.[1] The International Military Tribunal at Nuremberg (which articulated the Nuremberg principles—that there are such things as war crimes and crimes against humanity, that individuals can be held criminally responsible for committing them, and that "I was just obeying orders" is no excuse[2]—that serve as a basis for international criminal law, and in which judges from the four Allied powers presided) was followed by 12 "subsequent trials," each presided over solely by American judges.[3]

The first of the subsequent trials was the "Doctors' Trial," a trial of 23 physicians and scientists for murderous and torturous experiments conducted in the Nazi concentration camps.[4] The most infamous of these were the high-altitude experiments and the freezing experiments, both of which resulted in the planned death of the research subjects and both of which were conducted with the rationale that the results would help German pilots survive and so the experiments were necessary for the good of the survival of German society.[5] The American judges rejected the defense that the experiments were necessary and acceptable in wartime. In their final judgment, condemning the experiments and most of the defendants, seven of whom were hanged, the court articulated what is now known as the Nuremberg Code.[6] This ten-point code governing human experimentation was articulated by the American judges and based on what they had heard at trial, including the arguments of American prosecutors and the American physicians who served in the roles of consultant (Leo Alexander) and expert witness (Andrew Ivy) for the prosecution.[7]

Reaching the conclusion that American bioethics was born at the Nuremberg Doctors' Trial after exploring the post-World War II history of bioethics and human rights evokes T.S. Eliot's fabled lines from *Little Gidding*:

> We shall not cease from exploration
> And the end of all our exploring
> Will be to arrive where we started
> And know the place for the first time.
> (T.S. Eliot, excerpt from "Little Gidding," no. 4 of "Four Quartets")

It is coincidental, but fitting nonetheless, that T.S. Eliot composed these lines during World War II when he was a night fire-watcher during the fire bombings of London. World War II was the crucible in which both human rights and bioethics were forged, and they have been related by blood ever since. As I have already suggested, recognizing and nourishing this birth relationship will permit American bioethics to break free from its focus, if not obsession, with the doctor-patient relationship and medical technology and to cross our own border to become a global force for health and human rights—not as an imperialistic project, but to learn from and work with other cultures, countries, and activists. It may also help us answer another question Professor Wiesel posed after learning of contemporary torture at Abu Ghraib and Guantanamo Bay: why the "shameful torture to which Muslim prisoners were subjected by American soldiers [has not] been condemned by legal professionals and military doctors alike?"[8]

NAZI DOCTORS AND AMERICAN BIOETHICS

Although the World War II origin of American bioethics is easier to see at the beginning of the twenty-first century, mainstream bioethics historians, while acknowledging the Nuremberg Doctors' Trial and the Nuremberg Code as important historical events, continue to prefer seeing American bioethics as a 1960s–1970s response to medical paternalism made more powerful by an increasing volume of medical research and the development of new medical technology, especially organ transplantation and mechanical ventilation.[9] Nuremberg is seen as an important event, but one that had no immediate impact on medical ethics. One of the main reasons for this has been an active program to bury the Nazi doctor past and distance American medicine and American bioethics from Nazi medicine for fear it would be somehow tarnished by it.[10] The best known example is probably Henry Beecher—an anesthesiologist sometimes himself credited with getting American bioethics started with his 1966 article in the *New England Journal of Medicine* that catalogued a series of unethical experiments conducted at major U.S. research institutions long after the promulgation of the Nuremberg Code.[11]

Beecher was a leader in drafting the World Medical Association's (an organization formed in London at the end of 1946 just as the Doctors' Trial was getting underway) Helsinki Declaration on human research[12]—which many saw as a way to "save" medical research from becoming dominated by the "overly rigid" Nuremberg Code.[13] Nuremberg was considered overly rigid because of what psychiatrist Jay Katz consistently highlighted and praised about it, its "uncompromising language to protect the inviolability of subjects of research."[14] By putting the liberty and welfare of research subjects over the promise of medical progress, the Nuremberg judges sought to put the interests of individual humans over the interests of society in medical progress. But medical progress has consistently won out over the consent principle in the real world.[15] The "Belmont Report" of 1979, probably the most cited government-sponsored statement of research ethics, for example, begins with an opening paragraph about the Nuremberg Code, but then quickly asserts that it is "often inadequate to cover complex situations," like research on children and the mentally disabled.[16]

Nuremberg was also on the minds of Daniel Callahan and the founders of the Hastings Center, and they held a major program on its implications for bioethics.[17] But, as described by Arthur Caplan (who himself sponsored a similar program a decade later in 1989), there were many reasons for American bioethics to suppress its birth, most notably the sheer unprecedented scale of immorality of the Nazi doctors, and potential guilt by association, especially in the research enterprise.[18] But suppression did not prevent Caplan from concluding that "bioethics was born from the ashes of the Holocaust."[19]

The source of American bioethics can also be read in the biographies of almost all of the founders of American bioethics and its current leaders. The history of American bioethics is rooted in the Nazi concentration camps in another way as well. Historians are correct to see American bioethics in the late 1960s and early 1970s as fundamentally a reaction to powerful new medical technologies in the hands of medical paternalists who disregarded the wishes of their patients. Thus, the major strategy to combat this unaccountable power was to empower patients with the doctrine of informed consent (sometimes called autonomy and put under the broader rubric of respect for persons). This is perfectly reasonable. But it is unreasonable to want to distance yourself so much from your origins to miss the fact that Nazi physicians who performed experiments in the concentration camps did so in an impersonal, industrial manner on people they saw as subhuman and were unaccountable in the exercise of their power over their subjects. The first response of the American judges to the horror of the Nazi doctors was to articulate, in the first precept of the Nuremberg Code, the doctrine of informed consent.[20] The modern doctrine of informed consent was not born of either U.S. health law in 1972 or American bioethics shortly thereafter, but at Nuremberg in 1947. The importance of the Nuremberg Code and its requirement of informed consent for lawful experimentation were underlined by the U.S. Court of Appeals for the Second Circuit in early 2009. The court ruled that the families of victims of a Pfizer-sponsored experiment on children in Kano, Nigeria, in 1996 could sue Pfizer in the United States for failure to obtain informed consent because the Nuremberg Code's consent principle is now part of the "law of nations."[21]

HEALTH LAW, BIOETHICS, AND HUMAN RIGHTS

Misidentifying the birth of bioethics has also helped us to misidentify the birth of its primary doctrine, informed consent. American bioethicists have spent so much energy denying their origins that they have produced a misleading account of their central doctrine as well. The American judges at Nuremberg were also comfortable crossing borders, especially the border between American medical ethics (what we now know as bioethics) and international human rights law.

As in any organic whole, the boundaries between the interrelated fields of health law, bioethics, and human rights are easily crossed. The collapsing of other boundaries in human rights discourse suggests how a more integrative model might be built. In the brief history of human rights, for example, there have been three great divisions—all of which have been breached (although attempts to police these borders persist). These are the divisions between positive and negative rights, between public and private actors, and between state internal affairs and matters of universal concern.[22]

The positive/negative distinction has been seen more and more as a difference in degree rather than kind. This is because at least some positive government action is required even to ensure so-called negative rights, such as the right to be left alone, the right to vote, freedom of speech, and the right to trial by jury. All of these negative rights actually require the government to do something positive—such as setting up a police and court system and making legal counsel available to the accused. Of course, in the arena of positive rights, like the right to food, shelter, jobs, and health care, governments will be required to expend more resources (many more than for "negative" rights) to fulfill these rights. But resources will have to be expended to fulfill both types.

In the language of contemporary human rights, governments don't simply have the obligation to act or not to act, but rather have obligations regarding all rights to *respect* rights themselves, to *protect* citizens in the exercise of rights, and to *promote* and *fulfill* rights. Of course, not all governments can fulfill economic rights immediately because of financial constraints, and international law suggests that governments must work toward the "progressive realization" of these rights within the limits of their resources. Some governments may be so limited in their resources that they may require assistance from the world community, and the novel but powerful "right to development" speaks to the obligations of the world community to provide that assistance, as does the United Nations' (UN) Millennium Declaration.[23]

A similar analysis can be made of the distinction between private and public. Individuals cannot be free to commit crimes in the privacy of their homes; the law has jurisdiction in both the public and private sphere. And although international law has traditionally focused solely on the relationships between governments (and between a government and its people), private actors, like transnational corporations, have more recently been seen as having so many direct relationships with governments, who often act explicitly to protect their interests, that they should be seen as a fit subject for international human rights. Similarly, although historically the boundary of a country protected it from interference with its "internal affairs," the world today will not always simply stand by and watch as countries engage in massive human rights abuses (as the world did in Rwanda and continues to do in the Sudan), but may rather, as in South Africa, intervene to try to prevent major human rights abuses.

Entirely new entities, termed nongovernmental organizations, or simply NGOs, have sprung up and become the leading forces for change in the world. A notable health-related example is *Médecins sans Frontières* (MSF), a humanitarian-human rights organization founded on the belief that human rights transcend national borders, and thus, human rights workers cannot be constrained by borders, but should cross them when necessary. As Renée Fox describes it, over the years the *le droit d'ingerence* (the right to interfere) has been displaced with an even more activist *le devoir d'ingerence* (the duty to interfere).[24] This concept understands human rights as universal and sees globalization as a potential force for good. MSF expands medical ethics to include physician action to protect human rights, blending these two fields and treating the law that protects government territorial boundaries as subordinate to the requirements of protecting human rights. In this regard, MSF itself can be seen as one of the first health and human rights fruits of our human rights tree. Other notable physician NGOs that have taken the lead in adopting a human rights framework for their work include Physicians for Human Rights,[25] Global Lawyers and Physicians,[26] and perhaps most notably, the British Medical Association.[27]

THE UNIVERSAL DECLARATION OF HUMAN RIGHTS

Globally, boundaries are being breached by ideas, communication systems, and economics even as the world paradoxically splinters into more and more countries. Nonetheless, as daunting and discouraging as many contemporary challenges are, especially those related to global terrorism, the international research in genetic engineering and human cloning,[28] and the provision of basic health care to everyone, the Universal Declaration of Human Rights (UDHR) really does provide the world with an agenda and a philosophy.[29] The centrality of the UDHR to bioethics is well recognized internationally, for example, in United Nations Education, Science and Cultural Organization's (UNESCO) new "Declaration on Bioethics and Human Rights."[30]

The Cold War is well recognized as the force that prevented a single treaty from incorporating the principles of the UDHR and, instead, two separate treaties were drafted—one for civil and political rights and the other for economic, social, and cultural rights, reflecting the East-West divisions of government ideologies during the 1950s and 1960s. This separation was political and artificial, and it is now well recognized that economic and social (positive) and civil and political (negative) rights are interconnected and interrelated and that human beings need both to enable human flourishing. Less well recognized is that it was also the Cold War that prevented or at least slowed the development of American bioethics that originated with the Nuremberg Code. Because of fear of the Soviet Union, the United States acted much more pragmatically than principled in not only performing research, especially in the area of radiation research, not permissible under the Nuremberg Code (and thus required suppression or marginalization of the Code), but also actively recruited Nazi scientists and physicians to continue their research in the United States under U.S. military auspices.

The world's one remaining superpower and empire builder, the United States, has yet to enthusiastically embrace the UDHR—even though it was drafted under the able direction of Eleanor Roosevelt[31]—and has turned itself into an object of fear and distrust around the world in the wake of our "preemptive war" in Iraq.[32] But our government's attempt to ignore the precepts of the UDHR cannot ultimately prevail, and ignoring its political and civil precepts is fundamentally anti-American. The same can be said of the use of torture, a war crime and crime against humanity we prosecuted at Nuremberg.

Nonetheless, attempts to regain America's moral status as a proponent of human rights and its legal status as a country that follows the rule of law are not just hallmarks of the new Obama administration. For example, in late 2005 the Senate voted 90 to 9, over the objections of President George W. Bush, to both affirm our commitment to the UN Convention Against Torture and explicitly outlaw "cruel, inhuman, or degrading treatment or punishment" of anyone in the custody or control of the U.S. government.[33] The chief sponsor of this legislation, Senator John McCain, began his floor speech on his amendment to the Department of Defense Appropriations bill by saying,

> [L]et me first review the history. The Universal Declaration of Human Rights, adopted in 1948, states simply, "No one shall be subjected to cruel, inhuman or degrading treatment or punishment." The International Covenant on Civil and Political Rights, to which the United States is a signatory, states the same.[34]

Few Americans, I'm sure, ever thought that their government would condone and practice torture and inhuman and degrading treatment, let alone publicly justify torture as necessary for national security. Nonetheless, the Bush administration's torture position, as outlined in secret Department of Justice legal memos prepared for the CIA and released in April 2009, is consistent with a view of American pragmatism that says there are times when principles must be ignored to produce a result that is highly desired, and, when fighting evil (whether in war, or in a war against disease and death), it is acceptable to use an inherently evil means.[35] This justification for committing war crimes and crimes against humanity was, of course, rejected at Nuremberg by the United States.

McCain did not highlight the role and participation of physicians in torture and "aggressive interrogation" (nor has American bioethics had anything to say about either the war on terror or the role of physicians in it), but had he focused on physicians and medical ethics, he could have said even more about the Universal Declaration of Human Rights and the subsequent Covenant on Civil and Political Rights.[36] He could have noted that in adopting language for the 1958 Covenant, a treaty that the United States signed and that came into force in 1966, the Nuremberg Doctors' Trial was front and center on the minds of the drafters. The drafters added a second sentence to the original text of Article 5 of the UDHR, which they explained was added "in order to prevent the recurrence of atrocities such as those which had been committed in Nazi concentration camps during the Second World

War."[37] The two-sentence provision of Article 7 of the International Covenant on Civil and Political Rights reads in its entirety as follows:

> No one shall be subjected to torture or to cruel, inhuman or degrading treatment or punishment. In particular, no one shall be subjected without his free consent to medical or scientific experimentation.[38]

The drafting of the treaty on civil and political rights and its result, of course, means that Nuremberg and its consent principle were taken very seriously by the international law community in the 1950s.

Commenting on his experiences with top Bush administration lawyers who signed off on or wrote memorandums justifying torture, Alberto Mora, general counsel to the U.S. Navy from 2001 to 2006, said, "I wondered if they were even familiar with the Nuremberg trials—or with the laws of war, or with the Geneva [C]onventions."[39] He was right to wonder. In retrospect it appears that many of these lawyers did know about Nuremberg and the international laws of war, but just didn't care—their working hypothesis was that all that mattered was U.S. law and that the United States had no obligation to follow the international law it had helped to make. This is astonishing, as it doesn't take a high-power lawyer to understand that no individual country can unilaterally change international law. War crimes remain war crimes even if a country authorizes its agents to murder or torture. Nor does a "new kind of war" suspend the laws of war.[40] Churchill made this point shortly after World War II when he was writing his memoirs. In them, he describes what he calls "a terrible decision of policy adopted by Hitler" on June 14, 1941, at the outset of Germany's war with the Soviet Union.[41] Speaking to Generals Franz Halder and Wilhelm Keitel, Hitler said this war was "an entirely new kind of war" and thus the accepted international laws of war would not apply. In Halder's words,

> The Fuhrer stated that the methods used in the war against the Russians will have to be different from those used in the war against the West. . . . He stated that since the Russians were not signatories to The Hague Convention [precursor to the Geneva Conventions], the treatment of their prisoners of war does not have to follow the Articles of the Convention.[42]

The point is not that President Bush was acting like Hitler when he suspended the Geneva Conventions for the war on terror; the point is Alberto Mora's point: the president and his advisers seemingly knew nothing of the history of World War II, or certainly they would not have modeled their actions on Hitler—especially while declaring in public that they were acting like Churchill.[43] Similarly, the persistent and prolonged force-feeding of hunger strikers at Guantanamo by strapping them into "restraint chairs," which are the functional equivalent of straight-jackets, can be viewed not as "saving lives" but as human experimentation without consent. This view, which would likely seem reasonable to a reviewing court, and certainly would have to the judges at the Doctors' Trial, seems not to have even occurred to the military medical personnel at Guantanamo.[44]

It is, I think, the ability to see enemies as less than human that permits us to engage in inhuman acts without acknowledging, at least to ourselves, that this is what we are doing. This was also the primary theory behind Nazi eugenics, that is, that there were certain lives that were not worth living and that it was, therefore, justifiable to sterilize and ultimately to euthanize those who fit this category. Applied to large segments of the population, eugenics has a racist rationale. Because of the Nazis and the Holocaust, it seems unlikely that a concentration camp-based racist eugenics is likely to recur. Contemporary genetics and genetic screening seem much more benign. But the language is uncomfortably similar. An example is provided by James Watson, a fellow author in this book.

EQUALITY AND GENOMICS AND THE RISK OF GENISM

Equality based on human dignity is at the core of a human rights approach to health. For example, a country's obligation to "respect" and "protect" the right to health requires governments to "refrain from denying or limiting equal access to all persons" and to "ensuring equal access to health care."[45] The new genetics can be seen as scientific validation of human equality in that it demonstrates that we all share substantially identical genomes; but it can also be used to foster prejudice and discrimination and, thus, to undercut the right to health. This human tendency to create divisions, which I'm sure at least some people would describe as genetic, is well illustrated by an incident in late 2007 when James Watson, the codiscoverer of the structure of DNA, scandalized the world when he told a British newspaper, "I'm inherently gloomy about the prospect of Africa because all our social policies are based on the fact that their intelligence is the same as ours, whereas all the testing says not really."[46]

Watson later apologized and acknowledged that there is no scientific evidence to support his statement about differences in intelligence among races.[47] *Nature* magazine editorialized that Watson's remarks were "rightly . . . deemed beyond the pale," but also warned, "There will be important debates in the future as we gain a fuller understanding of the influence of genetics on human attributes and behavior. Crass comments by Nobel laureates undermine our very ability to debate such issues, and thus damage science itself."[48]

Our superficial perceptions of each other have often fostered racism in the past. Simply defined, racism is "the theory that distinctive human characteristics and abilities are determined by race."[49] The hunt for genes, especially in groups identified by racial classifications, could lead to "genism" (a term not yet officially recognized, but one I would define as "the theory that distinctive human characteristics and abilities are determined by genes") based on DNA sequence characteristics with resulting discrimination as pernicious as racism. Watson's remark was not one of an old-time racist, but of a new-style "genist."

It is true that "we are all Africans under the skin."[50] It is also true, however, that if we decide to search for genetic differences in the 0.5 percent of our DNA that is different, we will find them and use them against each other. Philosopher Eric Juengst put it well: "No matter how great the potential of population genomics to show our interconnections, if it begins by describing our differences it will inevitably produce scientific wedges to hammer into the social cracks that already divide us."[51]

Preventing genism from taking over where racism left off by substituting molecular differences for skin color differences will not be easy. Two actions, however, seem necessary. First, genetic privacy must be protected.[52] No one's genes should be analyzed without express authorization, and of course, no "genetic identity cards" should be permitted. Second, pseudoscientific projects that purport to identify genetic differences between "races" should be rejected.

Visions of the Future

The future that many American bioethicists, notably those on President Bush's Council of Bioethics, continue to worry about is Huxley's *Brave New World*—a world in which humans would be commodified and stratified and would give up all of their dignity and self-respect for security and recreational drugs and sex.[53] It was a world of humans reduced to animal status. Preventing this vision from becoming a reality is a reasonable goal. But exclusive concentration on a *Brave New World* vision and an embryocentric view of ethics energized by antiabortion sentiments, is not so much about bioethics as biopolitics, specifically President Bush's limitations on federal funding for human embryonic stem cell research to placate his Christian fundamentalist base. Bioethics is important in U.S. politics, just as morality is important in lawmaking. But when bioethics is used primarily to serve an ideological, domestic political agenda rather than helping to develop a global ethic, it is of little use to anyone other than narrow interest groups.

Making bioethics the servant of domestic politics also narrows its focus such that it is incapable of responding to or affecting a changing world, one envisioned more accurately in Orwell's *1984*:[54] a post-9/11 world dominated by military dictatorships kept in power by fear induced by "perpetual war," debasement of language (doublethink), and constant rewriting of history. Guantanamo prison camp is emblematic of our *1984* syndrome, and the fact that bioethicists have had almost nothing to say about the role of physicians there in "aggressive interrogation" and force-feeding (termed "assisted feeding" in doublespeak) hunger strikers demonstrates its real-world limitations. What seems evident is that human rights activists are more likely to provide nourishment to the human rights tree than bioethics theorists or health law scholars. Nonetheless, having practitioners of these interrelated fields working together has the potential to radically increase their impact on the real world—and for the better. This is why, rather than abandoning health law and bioethics for human rights, we renamed our department in the Boston University School of Public Health (formerly the Health Law Department) the Department of Health Law, Bioethics and Human Rights.

Salman Rushdie also had border crossings on his mind when he reflected on the meaning of 9/11 in his collection entitled *Step Across This Line*.[55] He ends his reflections by noting that "we are living, I believe, in a frontier time, one of the great hinge periods in human history, in which great changes are coming about at great speed." On the plus side, he lists the end of the Cold War, the Internet, and the completion of the Human Genome Project; on the minus side, a "new kind of war against new kinds of enemies fighting with terrible new weapons." The changes we will adopt are not preordained, and Rushdie quite properly notes that "the frontier both shapes our character and tests our mettle." He is also right to wonder whether, as we stand on this frontier, we will regress into barbarism ourselves or "as custodians of freedom and the occupants of the privileged lands of plenty, go on trying to increase freedom and decrease injustice?" A globalized American bioethics, infused with human rights, would have to pursue global justice.

NOTES

This chapter is adapted and updated from the final chapter of George J. Annas, *American Bioethics: Crossing Human Rights and Health Law Boundaries* (Oxford University Press, New York, 2005).

1. Evelyne Shuster, "Fifty Years Later: The Significance of the Nuremberg Code," *New England Journal of Medicine* (1997): 1439–1440; and Henry Steiner and Philip Alston, *International Human Rights in Context: Law, Politics, Morals*, 2nd ed. (New York: Oxford University Press, 2000), 56–135.
2. Robert F. Drinan, "The Nuremberg Principles in International Law," in *The Nazi Doctors and the Nuremberg Code: Human Rights in Human Experimentation*, ed. George J. Annas and Michael A. Grodin (New York: Oxford University Press, 1993), 174, 175.
3. Shuster, "Fifty Years Later," 1437.
4. Ibid.
5. Telford Taylor, Opening Statement of the Prosecution, December 9, 1946, in Annas and Grodin, *The Nazi Doctors and the Nuremberg Code*, 67, 71–75; and Arthur L. Caplan, "How Did Medicine Go So Wrong?" in *When Medicine Went Mad: Bioethics and the Holocaust*, ed. Arthur L. Caplan (Clifton, NJ: Humana Press, 1992), 71–77.
6. Michael A. Grodin et al., "Medicine and Human Rights: A Proposal for International Action," *Hastings Center Report* 23, no. 8 (1993): 8–9.
7. For a history of Leo Alexander and Andrew Ivy, see Shuster, "Fifty Years Later," 1437; and see generally Ulf Schmidt, *Justice at Nuremberg: Leo Alexander and the Nazi Doctors' Trial* (New York: Palgrave McMillan, 2004).
8. Elie Wiesel, "Without Conscience," *New England Journal of Medicine* 352 (2005): 1511–1513.
9. See, e.g., Albert R. Jonsen, *The Birth of Bioethics* (New York: Oxford University Press, 1998), 134; cf. David Rothman, *Strangers at the Bedside*, 2nd ed. (Hawthorne, NY: Walter de Gruyter, 2003), 62.
10. Caplan, *When Medicine Went Mad*, 78–79.
11. Henry K. Beecher, "Ethics and Clinical Research," *New England Journal of Medicine* 274 (1966): 1354–1360.
12. World Medical Association Declaration of Helsinki: Ethical Principles for Medical Research Involving Human Subjects, http://ohsr.od.nih.gov/guidelines/helsinki.html.
13. Sir William Refshauge, "The Place for International Standards in Conducting Research on Humans," *Bulletin of the World Health Organization* 55, no. 2 (1977): 85–92.
14. Jay Katz, "Human Sacrifice and Human Experimentation: Reflections at Nuremberg," *Yale Law School Occasional Papers, Paper 5.* (October 25, 1996), http://lsr.nellco.org/yale/ylsop/papers/5.
15. Renee Fox, "Medical Humanitarianism and Human Rights: Reflections on Doctors Without Borders and Doctors of the World," reprinted in *Health and Human Rights: A Reader*, ed. Jonathan Mann et al. (New York: Routledge, 1999), 417, 433; Tom L. Beauchamp, "Does Ethical Theory Have a Future in Bioethics?" *Journal of Law, Medicine and Ethics*

32 (2004): 209, 211; Renee Fox and Judith Swazey, "Medical Morality is not Bioethics: Medical Ethics in China and the United States," *Perspectives in Biology and Medicine* 27 (1984): 336–360; and Renee C. Fox and Judith P. Swazey, "Leaving the Field," *Hastings Center Report* 22 (1992): 9, 15.

16. National Commission for the Protection of Human Subjects of Biomedical and Behavioral Research, "The Belmont Report: Ethical Principles and Guidelines for the Protection of Human Subjects of Research," April 18, 1979, http://www.emerson.edu/graduate_studies/upload/belmontreport.pdf.
17. Daniel Callahan et al., "Special Supplement: Biomedical Ethics in the Shadow of Nazism," *Hastings Center Report* 6, no. 1 (1976): 1–19.
18. Caplan, *When Medicine Went Mad*, 78–79.
19. George J. Annas quoting Arthur Caplan, "American Bioethics and Human Rights: The End of All Our Exploring," *Journal of Law, Medicine and Ethics* 32 (2004): 658–659.
20. The Nuremberg Code, *Trials of War Criminals before the Nuremberg Military Tribunals under Control Council Law, No. 10* at 89, October 1946–April 1949, http://www.ushmm.org/research/doctors/Nuremberg_Code.htm.
21. Abdullahi v. Pfizer, 2009 U.S. App. LEXIS 1768 (2d Cir. 2009); and George J. Annas, "Globalized Clinical Trials and Informed Consent," *New England Journal of Medicine* 360 (2009): 2051–2053.
22. Steiner and Alston, *International Human Rights in Context*, 56–135.
23. United Nations General Assembly, Resolution 55/2, "Millennium Declaration," September 8, 2000, http://www.un.org/millennium/declaration/ares552e.htm.
24. Renee Fox, "Medical Humanitarianism and Human Rights," 420–421.
25. Physicians for Human Rights, History and Mission, http://physiciansforhumanrights.org/about/mission.html.
26. Global Lawyers and Physicians Home Page, http://www.glphr.org. My colleague, Michael Grodin, and I followed up our conference on the fiftieth anniversary of the Nuremberg Code at the Holocaust Memorial Museum by founding our own physician NGO—but combining it with lawyers as well: Global Lawyers and Physicians. The basic concept behind this NGO is that the professions of law and medicine are both inherently transnational and that by working together they can be a much more powerful force for promoting human rights than either profession can be working by itself.
27. British Medical Association, Medical Ethics, http://www.bma.org.uk/ethics/index.jsp.
28. George J. Annas, Lori B. Andrews, and Rosario M. Isasi, "Protecting the Endangered Human: Toward an International Treaty Prohibiting Cloning and Inheritable Alterations," *Journal of Law, Medicine and Ethics* 28 (2002): 151–178.
29. United Nations, The Universal Declaration of Human Rights, December 10, 1948, http://www.un.org/Overview/rights.html.
30. United Nations Education, Science and Cultural Council, International Bioethics Commission (IBC), *Report of the IBC on the Possibility of Elaborating a Universal Instrument on Bioethics*, prepared by Giovanni Berlinguer and Leonardo De Castro, U.N. Doc. SHS/EST/02/CIB-9/5, June 13, 2003.
31. Mary Anne Glendon, *A World Made New: Eleanor Roosevelt and the Universal Declaration of Human Rights* (New York: Random House, 2001), 79–98.
32. For a defense of the Bush Doctrine, see Philip Bobbitt, *Terror and Consent: The Wars for the Twenty-First Century* (New York: Knopf, 2008), 429–451.
33. David Rogers, "Senate in 90–9 Vote Passes Bill Seeking Clearer Detainee Rules," Wall Street Journal, October 6, 2005.
34. Congressional Record—Senate, S11063, 151, statement of Senator John McCain regarding Amendment No. 1977 of the *Department of Defense Appropriations Act, 2006*, October 5, 2005.

35. Jack Goldsmith, *The Terror Presidency: Law and Judgment Inside the Bush Administration* (New York: W.W. Norton & Co., 2007); and Jane Mayer, *The Dark Side: The Inside Story of How the War on Terror Turned into a War on American Ideals* (New York: Random House, 2008).
36. UN General Assembly, Official Records, Supplement 16, *International Covenant on Civil and Political Rights*, Office of the United Nations High Commissioner for Human Rights, in pursuance of UN General Assembly Resolution 2200A (XXI), Article 49, December 16, 1966.
37. Article Seven, draft *Covenant on Civil and Political Rights*, adopted by the Third Committee of the General Assembly of the United Nations, 1958 as cited in *Clinical Investigation in Medicine: Legal, Ethical and Moral Aspects*, ed. Irving Ladimer and Roger W. Newman (Law-Medicine Research Institute, Boston University, 1963), 162.
38. Ibid.
39. Jane Mayer, "The Memo: How an Internal Effort to Ban the Abuse and Torture of Detainees was Thwarted," *New Yorker*, February 27, 2006, 32, quoting Alberto Mora.
40. Jane Mayer, *The Dark Side.*
41. Winston S. Churchill, *The Second World War: The Grand Alliance* (Boston: Houghton Mifflin, 1950), 368.
42. Ibid.
43. George J. Annas, "Human Rights Outlaws: Nuremberg, Geneva, and the Global War on Terror," *Boston University Law Review* 87 (2007): 427, 430.
44. Ibid., 445–447.
45. M. Magdalena Sepúlveda, *The Nature of the Obligations under the International Covenant on Economic, Social and Cultural Rights* (Antwerpen: Intersentia nv, 2003), 204.
46. Charlotte Hunt Grubbs, "The Elementary DNA of Dr. Watson," *The Times* (London), October 14, 2007.
47. Rajeev Syal, "Nobel Scientist Who Sparked Row Says Sorry—I Didn't Mean It," *The Times* (London), October 19, 2007.
48. Editorial, "Watson's Folly," *Nature* 449 (2007): 948; and John Schwartz, "DNA Pioneer's Genome Blurs Race Line," *New York Times*, December 12, 2007.
49. *Oxford English Dictionary*, 2nd. ed. s.v. "racism."
50. Spencer Wells, interview by Syed Firdaus Ashraf, The Rediff Interviews, November 27, 2002, http://www.rediff.com/news/2002/nov/27inter.htm.
51. Eric Juengst, "Groups as Gatekeeper to Genomic Research: Conceptually Confusing, Morally Hazardous, and Practically Useless," *Kennedy Institute of Ethics Journal* 8 (1998): 183–200.
52. George J. Annas, Patricia Roche, and Robert C. Green, "GINA, Genism and Civil Rights," *Bioethics* 22, no. 7 (2008), ii.
53. Aldous Huxley, *Brave New World* (London: Chath and Windus, 1932).
54. George Orwell, *Nineteen Eighty-Four* (London: Secker and Warburg, 1949).
55. Salman Rushdie, *Step Across This Line* (New York: Random House, 2002).

CHAPTER 10

A More Perfect Human: The Promise and the Peril of Modern Science

Leon R. Kass

As nearly everyone appreciates, we live near the beginning of the golden age of biomedical science and technology. For the most part, we should be mightily glad that we do. We and our friends and loved ones are many times over the beneficiaries of its cures for diseases, prolongation of life, and amelioration of suffering, psychic and somatic. Since the latter third of the last century, most human beings living in technologically advanced countries have been living healthier and longer lives than even the most fortunate individuals in prior human history. Diphtheria, typhoid, and tuberculosis threaten us no longer; despite the lack of a definitive cure, half the people who are today treated for deadly cancers survive more than five years. The average American's life expectancy at birth has increased from 47 in 1900 to 78 in 2000, and millions are now living healthily into their eighties and nineties. Thanks to basic research in neuroscience and new psychotropic drugs, the scourge of major depression and other devastating mental illnesses are finally under effective attack. We have every reason to look forward eagerly to new discoveries and new medical blessings. Every one of us should be deeply grateful for the gifts of human ingenuity and for the devoted efforts of scientists, physicians, and entrepreneurs who have used these gifts to make those benefits possible.

Yet, notwithstanding these blessings, present and projected, we have also seen more than enough to make us concerned. For we recognize that the powers made possible by biomedical science can be used for nontherapeutic purposes, serving ends ranging from the frivolous and disquieting to the offensive and pernicious. These powers are available as instruments of bioterrorism (e.g., genetically engineered drug-resistant bacteria or drugs to obliterate memory); as agents of social control (e.g., drugs to tame rowdies and dissenters or fertility-blockers for welfare recipients); and as means of trying to improve or perfect our bodies and minds and those of our children (e.g., genetically engineered super-muscles or drugs to improve memory). Anticipating possible threats to our security, freedom, and even our very humanity, many people are increasingly worried about where biotechnology may be taking us. We are concerned not only about what others might do to us, but also about what we might do to ourselves. We are concerned that our society might be harmed and that we ourselves might be diminished, indeed, in ways that could undermine the highest and richest possibilities of human life.

Most of us cheerfully believe that we can both enjoy the promise and escape the perils of the coming biotechnological age. Most of us also believe that there is little connection between the promise and the peril, or between the humanistic aspirations that fuel the scientific enterprise and the deadly or dehumanizing uses to which new technologies might perversely be put. But a powerful challenge to our complacent opinions is provided by an important exhibit prepared by the United States Holocaust Memorial Museum, Deadly Medicine: Creating the Master Race.[1] This exhibit documents the abominable uses that the Nazis made of science and medicine. But, even more relevant for us, it also presents the scientific outlook on life and the aspiration to human perfectibility that the Nazis inherited and exploited, an outlook and an aspiration that dwell robustly in American cultural life today.

SCIENCE AS SALVATION: A CAUTIONARY TALE

The Deadly Medicine exhibit actually invites us to self-attention because of where it starts and how it is structured. The first of the exhibit's three parts, devoted to pre-Nazi Weimar eugenic ideas and practices, is entitled Science as Salvation. The next two parts display, respectively, The Biological State, through which those eugenic ideas were turned into Nazi racial hygiene (1933–1939), and, last, The Final Solution, through which Nazi racial hygienic practices became mass murderous (1939–1945). The exhibit thus locates the Nazi medical atrocities in the company of an idealistic science that preceded it, and it asks us to ponder whether there is any deep connection between the beauty of the Glass Man (see figure 10.1) and the night of broken glass and the horrors thereafter. The true power of the exhibit lies in the question it tacitly poses regarding the relation between the last phase and the first: what, if any, is the connection—not only historical but also *logical*—between the Final Solution and the disposition to look to science for salvation? How, if at all, are the optimistic dreams of building a more perfect human through science and medicine related to the actual building of death camps in which real human beings, deemed worthless and worse, were exterminated like so much vermin?

Now nothing in the exhibit suggests that the idealistic science of Weimar produced, or even *necessarily* led, to the Final Solution, though we learn here that the former sowed the seeds that were later used to grow the murderous fruit. And it is surely not the exhibit's even subliminal intention to suggest that noble science need elsewhere become, however unintentionally, the handmaid of bestiality or, further, that genetics or medicine or psychiatry—now as well as then—should come under suspicion of lending strength to deadly inhumanity. Yet, and on the other hand, the exhibit—to its great credit—does not allow us lovers of science and progress to rest

Figure 10.1 Glass Man display in the German Hygiene Museum of Dresden, 1930.
Source: Courtesy of the United States Holocaust Memorial Museum.

comfortably with the belief that the Nazis simply corrupted and perverted science or that their science wasn't really even science or that their nefarious purposes were worlds apart from the humanitarian aspirations of modern medicine. The exhibit compels us to consider whether the Nazi use of medical science might have been less a perversion of science, more a monstrously evil yet also logically fitting conclusion from certain dubious premises and attitudes in the scientific outlook itself and especially from the prevailing assumptions about the role of science in human affairs. Is there perhaps something wrong—even *deadly* wrong—in seeing science as our salvation? If so, then we might need to be on our guard when this siren song is sung *to us*, as it is today being sung by an ever larger, louder, and much more competent chorus.

To reach the ghastly result, the eugenic and perfectionist vision of Weimar had to be politicized by the Nazis, and in a most particular way. The project for the Final Solution depended decisively on the presence of a nearly omnicompetent totalitarian and tyrannical state, enforcing state-sponsored racial and ethnic hatreds, and assaulting the traditional teaching—both biblical and liberal democratic—of the irreducible and equal dignity of every human individual. God willing, we shall not see such a regime again.

Compassionate people like ourselves, who enjoy the protections of liberal democratic institutions, strong cultural prejudice favoring the individual against the collective, and, however much diminished, the invaluable Judeo-Christian belief in the sanctity of human life, can reasonably believe that "it cannot happen here," and I surely agree with this conclusion. But—and this is the first important point I wish to make—the explicitly *Nazi* elements of tyranny and race hatred are *not* absolutely necessary for producing a deadly medicine, even if it never again becomes a "holocaust." A free people, choosing for ourselves, can and very likely will produce similar deadly fruit from the same dangerous seeds, unless we are ever vigilant against the dangers. This chapter seeks to identify some of the deadly dangers that lurk in the seductive ideas and practices of science as salvation. The essence of the peril lies, ironically, in the zealous pursuit of the more perfect human.

A word of caution against a possible misunderstanding: although I shall be raising questions about the idea and practice of scientism, nothing that I shall say should be taken as "antiscience" or "antiscientists." The question before us is not the goodness of science and medicine as such, but the goodness of looking to science and medicine as the solution for the human condition, for the relief and salvation of man's estate.

The first two images from the Holocaust Museum's exhibit pose the problem and set the stage for all my reflections: the Glass Man and the photographs of World War I survivors (see figures 10.1 and 10.2).

The exhibit opens with the stunning Glass Man, first displayed at the German Hygiene Museum in Dresden in 1930. Though many of us have become familiar with transparent models of the human body—they are today widely marketed as science toys for school children—it is difficult to exaggerate the excitement that those original models created. For the first time, the common man could glimpse a life-like model of his insides, organ by organ, artery by artery, nerve by nerve, seeing with illuminated brilliance all the parts that make him run. Far from looking ashamed or diminished by this anatomizing invasion of his inner being, the Glass Man stands toweringly over us, fitly and proudly, with arms uplifted in a gesture of triumphant appeal for heavenly applause, a model of human perfection not to say apotheosis. Moreover, this perfect man clearly came not from the hand of God but from an even more perfect human, the scientific and medical visionary who would someday soon help humankind collectively achieve the healthful perfection here modeled in glass.

Make no mistake, this is serious business. For the Glass Man was willy-nilly the emblem of a new religion: in place of the God who became man, we have here the man become as god. In place of the suffering Christ, arms stretched in crucifixion, we have the impervious Glass Man, arms elevated in self-exaltation. And creatively

Figure 10.2 Maimed soldiers in Plötzensee near Berlin, summer 1916.
Source: Courtesy of USHMM.

behind the scene, in place of a God who it is said sent His son who would, through his own suffering, take away the sins of the world, we have the scientific savior who would take away the sin of suffering altogether. The Glass Man, *in loco crucifixis*, is the perfect icon for salvific science.

The dream of perfect health and fitness is, of course, quite ancient. Indeed, 300 years earlier it acquired an honored place among the goals of modern science, when its great founders, Francis Bacon and René Descartes, summoned humankind to the conquest of nature for the relief of man's estate. Mastery and possession of nature, Descartes announced,

> is desirable not only for the invention of an infinity of artifices which would enable us to enjoy, without any pain, the fruits of the earth and all the commodities to be found there, but also and principally for the conservation of health, which is without doubt the primary good and the foundation of all other goods in this life.

As the sequel makes clear, Descartes had his eye on goals grander than the mere absence of disease or even just ordinary bodily health:

> For even the mind is so dependent on the temperament and on the disposition of the organs of the body, that if it is possible to find some means that generally renders men *more wise* and *more capable than they have been up to now*, I believe that we must seek for it in medicine . . . . [W]e could be spared an infinity of diseases, of the body as well as of the mind, and *even also perhaps the enfeeblement of old age*, if we had enough knowledge of their causes and all the remedies which nature has provided us. [Emphasis added.][2]

But the ancient dream of perfect health and fitness had acquired a new prominence in Europe owing to the Great War, only recently ended. Images from the aftermath of that war, the second display of the exhibit, provide the counterpart to the Glass Man and in part explain his great social appeal (see figure 10.2).

Human deformity, from loss of limbs to loss of mind, came home from the war to Germany (and to all of Europe) by the tens and hundreds of thousands. The maimed and the enfeebled had rarely if ever been seen in such numbers—thanks, please note, to the great technological improvements for waging war—and the response of the German mind, humiliated in the War, did not take the most compassionate turn. On the contrary, fear and loathing of the deformed and the defective found abundant expression, as hatred of imperfection grew up with and encouraged the desire to imitate the perfection of the Glass Man.

In 1920, right after the war and well before the Nazi period, a distinguished jurist, Karl Binding, and a distinguished physician, Dr. Alfred Höche, both of them liberals, published a chilling booklet entitled "On Permitting (or Authorizing) the Destruction of Life Unworthy of Life *(Lebensunwerten lebens).*" Beginning modestly with a defense of the moral acceptability of suicide and assisted suicide, Binding and Höche moved cunningly to a defense of killing those whose miserable condition of body or mind calls for the "healing remedy" of premature death from the hand of medical science. Contemplating the battlefield strewn with thousands of dead youths and comparing this with the mental hospitals dedicated to the long-term care of the demented and the mentally ill, Binding comments,

> One will be deeply shaken by the strident clash between the sacrifice of the finest flower of humanity in its full measure on the one side, and by the meticulous care shown to existences which are not just absolutely worthless but even of negative value, on the other.[3]

And Dr. Höche ends his part of the booklet with a paean to the dawning of a new age:

> There was a time, now considered barbaric, in which eliminating those who were born unfit for life, or who later became so, was taken for granted. Then came the phase, continuing into the present, in which . . . preserving every existence, no matter how worthless, stood as the highest moral value. A new age will arrive—operating with a higher morality and with great sacrifice—which will actually give up the requirements of an exaggerated humanism and overvaluation of mere existence.[4]

A vigorous society, comprising only healthy and fit members, is more than justified in doing battle with the evils of deformity and disability by cleansing society of the disabled and the deformed themselves.

These twin goals—the positive goal of seeking perfection and the negative goal of removing imperfection—are, to repeat, nothing new. They are of ancient pedigree. Indeed, they inspire much of the good that we do in life, and not only in medicine. We pursue virtue or excellence; we stifle vice and correct mediocrity. We urge our children to be good and our societies to be better; we try to eliminate the deficiencies and evils to which they are subject. For Christians, the counsel of perfection is even a divine injunction—"Be ye therefore Perfect, even as your Father which is in heaven is Perfect"[5]—though the perfection Jesus had in mind, I am confident, cannot be pictured in glass.

To be sure, there is always a danger that we will turn our opposition to deficiency into a rejection of those who bear it, that, for example, the battle against ignorance or impairment will translate into a hatred for the ignorant or the impaired. Indeed, by calling them after their imperfection ("he is a paraplegic"; "she is a Down's"), we show our penchant for not distinguishing between the sin and the sinner. But this age-old tendency acquires a new character and a new momentum in an age that not only extols and exhorts to perfection, but also, and more important, gathers the scientific means to pursue it. So, just as the loathing of imperfection fuels the search for perfection, so the search for perfection makes imperfection all the more intolerable. Such is the inner meaning of science seen as salvation, informed by a new idea of human perfection that has, in the end, little patience with human frailty and disability. That attitude is once again gaining strength, and this time it comes with first-rate science and powerful and precise technique. And, in contrast to pre-World War II Germany, it speaks in the seductive voices of freedom, compassion, and self-improvement.

The technologies of interest touch all aspects of human life, from beginning through middle to end. Even as we stand but at the dawn of the new age ushered in by deciphering the entire human genome, we are already widely practicing genetic screening and prenatal and preimplantation genetic diagnosis, capable of identifying and rooting out the genetically unfit before they can be born. And advances in genetics, developmental biology, and neuroscience promise us all sorts of enhancements in human nature that would make us "better than well," in both body and mind. On the negative side, eliminating imperfections, the prime targets for prevention, correction, and elimination are mental retardation and mental illness, severe bodily deformity and disability, and, later in life, dementia, debility, and enfeeblement—serious imperfections all. On the positive side, promoting perfection, the prime targets for improvement are memory, muscularity, mood, temperament, intelligence, and—the Holy Grail—human finitude itself, to be ameliorated through the conquest of biological senescence.

Time does not permit a proper examination of even one of these promises. Instead, taking a more synoptic overview, but with some special attention to eugenics, I will briefly review some of the more obvious sources of danger, moral hazards that are especially difficult for us to recognize because our practices appear to be governed not by coercive state policy but by unconstrained free human choice. Aiming to present "the big picture," the better to awaken us from our complacency, I will paint with broad strokes. Accordingly, the demonstrations I offer are showings, appealing to our intuitions, not logical proofs, appealing to our analytic reason.

There are two sorts of moral hazards: dangerous practices and dangerous thinking.

Dangerous Practices: Negative and Positive Eugenics

In the exciting early days of the genetic revolution, the 1960s and 1970s, both positive and negative eugenic goals were enunciated with great gusto. Conferences were held with the bold titles of Genetics and the Future of Man. At one such meeting, the distinguished molecular biologist Robert Sinsheimer enthused that

> for the first time in all time a living creature understands its origins and can undertake to design its future.... [W]e can be the agent of transition to a wholly new path of evolution.[6]

About the same time, Nobel Laureate Joshua Lederberg saw in the prospect of human cloning an end to the rule of chance in human reproduction and the opportunity to perpetuate unaltered the genes of genius,[7] and many people looked forward to discovering whether a second Mozart might outdo the first. Also, about the same time, the president of the American Association for the Advancement of Science, the gentle geneticist Bentley Glass, enunciated a new right, "the right of every child to be born with a sound physical and mental constitution, based on a sound genotype." Looking ahead to the reproductive and genetic technologies that are today rapidly arriving, Glass proclaimed, "No parents will in that future time have a right to burden society with a malformed or a mentally incompetent child."[8]

Nowadays, we hear almost no such bold eugenics talk from mainstream scientists, though it is bruited about the margins by a small group of bio-prophets summoning us to a post-human future or a remaking of Eden. But eugenic vision and practice are gaining strength, all the more so because they grow out of sight behind the fig leaf of the doctrine of free choice. We are largely unaware that we have, as a society, already embraced the eugenic principle "Defectives shall not be born" because our practices are decentralized and because they operate not by coercion but by private reproductive choice. Genetic knowledge, we are told, is merely providing information and technique to enable people to make better decisions about their health or reproductive choices.

But our existing practices of genetic screening and prenatal diagnosis show that this claim is at best self-deceptive, at worst disingenuous. The choice to develop and practice genetic screening and the choices of which genes to target for testing have been made not by the public but by scientists—and not on liberty-enhancing but on eugenic grounds. Many practitioners of prenatal diagnosis refuse to do fetal genetic screening in the absence of a prior commitment from the pregnant woman to abort any afflicted fetus. Many pregnant women who wish *not* to know prenatal facts must withstand strong medical pressures for testing.

Practitioners of prenatal diagnosis, working today with but a fraction of the information soon to be available from the Human Genome Project, already screen for a long list of genetic diseases and abnormalities, from Down syndrome to dwarfism. Possession of any one of these defects, they believe, renders a prospective child unworthy of life. Persons who happen still to be born with these conditions, having somehow escaped the spreading net of detection and eugenic abortion, are increasingly regarded as "mistakes," as inferior human beings who should not have been born. Not long ago, at my own university, a physician making rounds with medical students stood over the bed of an intelligent, otherwise normal ten-year-old boy with spina bifida. "Were he to have been conceived today," the physician casually informed his entourage, "he would have been aborted." A woman I know with a child who has

Down syndrome is asked by total strangers, "Didn't you have an amnio?" The eugenic mentality is taking root, and we are subtly learning with the help of science to believe that there really are certain lives unworthy of being born.

Not surprisingly, in the face of these practical possibilities, prominent intellectuals are now providing justification for this view of life. The current journals of bioethics, no less, are filled with writings that sweetly sing the song of Binding and Höche, albeit without the menacing German accent. But not all are so reticent. Here, for example, are remarks from the writings of Peter Singer, DeCamp Professor of Bioethics in the University Center for Human Values at Princeton, on the question of killing infants with serious, yet manageable, diseases such as hemophilia:

> When the death of a disabled infant will lead to the birth of another infant with better prospects for a happy life, the total amount of happiness will be greater if the disabled infant is killed. The loss of a happy life for the first infant is outweighed by the gain of a happier life for the second [even if not yet born]. Therefore, if killing the hemophiliac infant has no adverse effect on others, according to the total view, it would be right to kill him.[9]

In a recent magazine interview, Singer was asked, "What about parents conceiving and giving birth to a child specifically to kill him, take his organs, and transplant them into their ill older children?" Singer replied, "It is difficult to warm to parents who can take such a detached view, [but] they're not doing something really wrong in itself." The interviewer then asked, "Is there anything wrong with a society in which children are bred for spare parts on a massive scale?" The Princeton professor of bioethics replied, "No."[10] Do not underestimate what it means for us that such coolly lethal opinions, regarded since 1945 as barbaric, are today again treated with seriousness, and that promoters of such opinions can occupy professorial chairs of ethics at places like Princeton.

Similar ideas and practices are coming into vogue at the other end of life. The practice of physician-assisted suicide and euthanasia has been legal in Holland for several decades and more recently in our own states of Oregon and Washington. After several quiet years, the campaign is heating up again, and several state legislatures are once again considering Oregon-type legislation. For a variety of reasons, an age of legalized euthanasia is likely to be soon upon us. Few are going to speak openly like Peter Singer about ending worthless lives. But like him, they will promote the deadly practice under the banner of autonomy and choice, graced with slogans of a dignified death—and, of course, utilitarian appeals to cut the costs of care.

To be sure, large familial and social difficulties of a mass geriatric society are already upon us, destined to become much more severe as the baby boomers enter upon their seniority—and, alas, senility. Although vast numbers of old people are today healthier and longer lived than ever before, the price many of them are paying for the extra decade of vigorous old age between 70 and 80 is often another decade between 80 and 90 of enfeeblement, debility, and dementia. Today, over 4.5 million Americans are afflicted with Alzheimer's disease, a number that is predicted to triple before midcentury. Thanks to our ability to treat acute illnesses and crises, roughly 40 percent of us can expect to spend roughly ten frail, enfeebled, and often demented years at the end of our lives, incapable of caring for ourselves in a world of fewer and fewer familial caregivers and, in most cases, without resources to purchase decent home or institutional care. Already we hear the dire statistics about the amount of health care costs spent futilely on the last six months of life. Already we hear the call for rationing, for not wasting resources on persons with "low quality of life"—a gentle Americanism for the German "life unworthy of life."

I do not minimize the ethical anguish that often confronts patients and families when loved ones linger on, their memories gone, their lives little resembling anything like the ones they enjoyed in their prime. But I still shudder when I hear the call for a technical quick solution to the need for long-term care, for I know what we have to fear when a shallow notion of death with dignity enlists deadly medical force to solve society's demographic and economic problems.

In Holland, we have seen a foretaste of the future. There the "right to die" has flowed down the slippery slope to its most radical meaning and then some: from a right to refuse treatment, to a right to control one's own dying, to a right to assistance in "becoming dead," to a right to voluntary euthanasia, to a right to be mercifully dispatched by one's doctor should *he* decide that you are "better off dead." The descent into unauthorized euthanasia is confirmed by official reports from Holland, with roughly a third of Dutch doctors, speaking under immunity, confessing that they have been practicing *non*voluntary euthanasia, without patient knowledge or consent, including on a significant number of patients who were mentally totally competent. In 2005 without anyone making a fuss, the Dutch issued a protocol for euthanizing severely ill newborns. If this can happen among the liberal and tolerant Dutch, are we so sure that it cannot happen here?

But the battle against imperfection by eliminating the imperfect is only part of our current eugenic story. Much more vigorous is our scientific and biotechnical quest for human improvement, for doing nature one better and making a more perfect human being. Embryos that are now screened for the presence of disease-causing genetic abnormalities may also soon be screened for the presence of certain desirable genetic traits, from perfect pitch to greater height to calmer temperament, even perhaps someday, to higher IQ. And although precise genetic engineering of designer babies seems to me to be pure science fiction, human cloning does hold out the prospect of trying to perpetuate tested superior genotypes. Genetic engineering of adults holds out the prospect of enhancing muscle bulk and performance. And, beyond genetic enhancement, psychoactive drugs are being developed to increase concentration, to erase troubling memories, or to alter personality. And there is active research to increase the maximum life expectancy, from hormone treatments to stem cell based, transplantable tissues to the ultimate weapon, the control of the genes that determine the rate of aging and the age of death. There is also active research on human-computer interactions, beginning with attempts to enable the deaf to hear and the blind to see, but culminating perhaps in computer implants in the brain that would enable us to download entire libraries at the click of mouse.

Apart from a few zealots such as the immortalists or the bionics boosters around *Wired Magazine*, most people exploring these prospects are not trying to build a superman or a post-human being. They are, by their own lights, just trying to enhance human performance by these more effective biotechnical means, offering a psychophysical route to human improvement that could supplement and extend the improvements we cultivate for ourselves through education or personal training. Yet, especially with large commercial interests hyping the benefits and creating new demands, there is no question but that such enhancements will be widely desired and used to satisfy the age-old personal human dreams of better children, superior performance, ageless bodies, and happy souls. They may even be enlisted to advance certain social goals—enabling soldiers and pilots to go without sleep, or schoolteachers and prison wardens to pacify the unruly.

We have only begun to consider the momentous ethical and social questions in store for us as we head down this road. A report from the President's Council on Bioethics, *Beyond Therapy: Biotechnology and the Pursuit of Happiness*, is an early effort to articulate our unease at these prospects.[11] We are right to be concerned about the meaning of pursuing venerable human goals by these "magical" technical means, just as we are right to be concerned about the wisdom of trying to transcend through technology the parameters of our given nature, the delicately balanced product of eons of gradual evolution.

Will human life really be better if we turn to biotechnology to fulfill our deepest human desires? Will those desires be properly satisfied? Will our enhanced activities really be *better*, and better *humanly*? There is an old expression, "To a man armed with a hammer, everything looks like a nail." To a society armed with biotechnology, the activities of human life may come to be seen in purely technical terms, and more amenable to improvement than they really are. Worse, like Midas, we may get more easily what we asked for only to realize it is vastly less than what we really wanted.

We want better children—but not by turning procreation into manufacture or by altering their brains to gain them an edge over their peers. We want to perform better in the activities of life—but not by becoming mere creatures of our chemists or by turning ourselves into bionic tools designed to win and achieve in inhuman ways. We want longer lives—but not at the cost of living carelessly or shallowly with diminished aspiration for living well, or by becoming people so obsessed with our own longevity that we care little about the next generations. We want to be happy—but not by means of drugs that give us happy feelings without the real loves, attachments, and achievements that are essential for true human flourishing.

The pursuit of these perfections, defined scientifically and obtained technologically, not only threatens to make us more intolerant of imperfection, both our own and our neighbor's. It also threatens to sell short the true prospects for human flourishing, which have always been found in love and friendship, work and play, art and science, song and dance, service and worship—not in chemically induced highs or bionic achievements. Our deepest longings are not for artificial contentment and factitious achievements, but for lives that are meaningful, connected, and humanly flourishing. In the absence of knowing what human flourishing is, there can be no perfecting of human beings or enhancement of human life. And no one can presume to judge any change in human nature or human activity to be an improvement—never mind a "perfection"—if they do not know what is humanly good.

DANGEROUS THINKING: SOULLESS SCIENTISM

The question we must, therefore, put to the human enhancers and the post-human futurists is this: what knowledge of the human good do you have that entitles you to gamble the human future on your hunches that these proposed alterations will in fact make us better or happier? It is a question that science and technology simply cannot answer, and worse, that our bio-prophets do not even think to ask. No danger we face in the coming age of biotechnology is greater than the danger of careless and shallow thinking.

If we are to avoid both the deadly and the dehumanizing results from our uses of biotechnology, we will need to be vigilant in our practices and resourceful in our thinking. Everything will depend on whether the technological disposition is allowed to proceed to its self-augmenting limits, or whether it can be restricted and brought under intellectual, spiritual, moral, and political rule. But on this front, I regret to say, the news is not encouraging. For the relevant intellectual, spiritual, and moral resources of our society, the legacy of civilizing traditions painfully acquired and long preserved, are taking a beating—not least because they are being called into question by the findings of modern science itself. The technologies present troublesome ethical dilemmas, but the underlying scientific notions—so we are told—call into question the very foundations of our ethics and our human self-understanding.

In the nineteenth and early twentieth centuries, the challenge came in the form of Darwinism and its seeming opposition to biblical religion, a battle initiated not so much by the scientists as by the beleaguered defenders of orthodoxy. In our own time, the challenge comes from molecular biology, behavioral genetics, neuroscience, and evolutionary psychology, fueled by their practitioners' overconfident belief in the sufficiency of their reductionist explanations of all vital and human phenomena and, in some cases, by an explicit intention to overthrow our venerable religious beliefs. Never mind "created in the image of God": what elevated *humanistic* view of human life or human goodness is defensible against the belief, trumpeted by biology's most public and prophetic voices, that man is just a collection of molecules, an accident on the stage of evolution, a freakish speck of mind in a mindless universe, fundamentally no different from other living—or even nonliving—things? What chance have our treasured ideas of freedom and dignity against the reductive notion of "the selfish gene" (or, for that matter, of "genes for altruism"), the belief that DNA is the essence of life, or the teaching that all human behavior and our rich inner life are rendered intelligible only in terms of neurochemistry or their contributions to species survival and reproductive success?

These transformations of moral outlook are, in fact, welcomed by many of our leading scientists and intellectuals. In 1997 the luminaries of the International Academy of Humanism—including biologists Francis Crick, Richard Dawkins, and E. O. Wilson and humanists Isaiah Berlin, W. V. Quine, and Kurt Vonnegut—issued a statement in defense of cloning research in higher mammals and human beings. Their reasons were revealing:

> What moral issues would human cloning raise? Some world religions teach that human beings are fundamentally different from other mammals—that humans have been

> imbued by a deity with immortal souls, giving them a value that cannot be compared to that of other living things. Human nature is held to be unique and sacred.... [But] as far as the scientific enterprise can determine,... [h]umanity's rich repertoire of thoughts, feelings, aspirations, and hopes seems to arise from electrochemical brain processes, not from an immaterial soul that operates in ways no instrument can discover.... Views of human nature rooted in humanity's tribal past ought not to be our primary criterion for making moral decisions about cloning.... [I]t would be a tragedy if ancient theological scruples should lead to a Luddite rejection of cloning.[12]

In order to justify ongoing research, these intellectuals were willing to shed not only traditional religious views but *any* view of human distinctiveness and special dignity, their own included. They fail to see that the scientific view of man they celebrate does more than insult our vanity. It undermines our self-conception as free, thoughtful, and responsible beings, worthy of respect because we alone among the animals have minds and hearts that aim far higher than the mere perpetuation of our genes. It undermines, as well, the beliefs that sustain our mores, practices, and institutions—including the practice of science itself.

The problem, in fact, lies less with the scientific findings themselves, more with the shallow scientistic "philosophy"—it is more properly called a faith—that recognizes no other truths but these and with the arrogant pronouncements of the bio-prophets. Here, for example, is the eminent psychologist Stephen Pinker railing against any appeal to the human soul:

> Unfortunately for that theory, brain science has shown that the mind is what the brain does. The supposedly immaterial soul can be bisected with a knife, altered by chemicals, turned on or off by electricity, and extinguished by a sharp blow or a lack of oxygen. Centuries ago it was unwise to ground morality on the dogma that the earth sat at the center of the universe. It is just as unwise today to ground it on dogmas about souls endowed by God.[13]

One hardly knows whether to be more impressed with the height of Pinker's arrogance or with the depth of his shallowness. Pinker is ignorant of the fact that "soul" need not be conceived as a "ghost in the machine" or as a separate "thing" that survives the body, but can be understood (à la Aristotle) to be the integrated powers of the naturally organic body. He has not pondered the relationship between "the brain" and the whole organism or puzzled over the difference between "the brain" of the *living* and "the brain" of the dead. He seems unaware of the significance of emergent properties, powers, and activities that do not reside in the materials of the organism but that emerge only when the materials are formed and organized in a particular way; he does not understand that this empowering organization of materials—the vital form—is not itself material. But Pinker speaks with the authority of science, and few are able and willing to dispute him on his own grounds.[14]

There is, of course, nothing novel about reductionism and materialism of the kind displayed here; these are doctrines with which Socrates contended long ago. What is new is that, as philosophies, they seem (to many people) to be vindicated by scientific advance. Here, in consequence, is perhaps the most pernicious result of our technological progress—more dehumanizing than any actual manipulation or technique, present or future: the erosion, perhaps the final erosion, of the idea of man as noble, dignified, precious, or godlike, and its replacement with a view of man, no less than of nature, as mere raw material for manipulation and homogenization.

Hence, our peculiar moral crisis. We are in turbulent seas without a landmark precisely because we adhere more and more to a scientific view of nature and of human life that both gives us enormous power and that, *at the same time*, denies every possibility of nonarbitrary standards for guiding its use. Though well equipped, we know not who we are or where we are going. We triumph over nature's unpredictabilities only to subject ourselves, tragically, to the still greater unpredictability of our capricious wills and our fickle opinions. Engineering the engineer as well as the engine, we race our train we know not where. Lacking any rich view of human flourishing, our pursuit of a more perfect human is at best chimerical. That we do not recognize our predicament is itself a tribute to the depth of our infatuation with scientific progress and our naive faith in the sufficiency of our humanitarian impulses.

Let me return, in conclusion, to look again at the Glass Man, seen now in the light of this discussion. It turns out that the Glass Man is in fact not transparent but opaque. It pretends to show us the innermost man, but it in truth renders his humanity permanently absent. Yes, we see the liver, the kidneys, and the colon, but we learn nothing about the soul. The mysterious character of the human person has not been explained; rather it has been ignored, nay, banished. The problem is not that anatomizing did not reveal the soul—no one thought it could. It is rather that anatomizing ignores and then denies the soul, denies the wholeness and the inner depth of the human being, even in the very act that seems for the first time to make it visible to him.

In one respect, however, the Glass Man reveals a permanent truth about the human being, ironically driving home a lesson that I do not believe that its makers meant to teach. When we look at the Glass Man's head and face, hoping to find evidence of the human soul within, what stares back at us is only the bony skull, universally the mark and symbol of death. Lurking beneath the outer surface of this godlike man is the truth about his vaunted perfection: alas, poor Yorick, death will be his fate, medicine or no medicine. And surely speaking better than his creators intended, the Glass Man's skull that betokens death for the individual human being betokens also the deadly consequence for a society that would pursue bodily perfection in ways that do not also join hands in solidarity with those who will never reach it.

The idealistic German scientists could not have known that the Glass Man was the harbinger of anything like the Final Solution. But they should have known that the biologizing and soulless account of human life that they were trumpeting is in fact always deadly to humanity—even if not one crematorium is built. A dehumanizing account of human life can all by itself produce a holocaust of the human spirit.

To keep human life human, we need first and foremost a more natural biology and anthropology, a robust account of the nature and meaning of our own humanity that will do justice to life as lived, with its high aspirations, deep longings, and rich possibilities for flourishing. In seeking such an account, we must draw on the wisdom of poets and philosophers and the insights of the great religious traditions. But one thing is clear: our penchant to think only in terms of the American ideals of freedom and equality will not be adequate to the task. In addition, we will need a robust account of *human dignity*, of our peculiarly human special standing, one that has been variously captured in notions that describe us as the "rational animal" or "made in the image of God" or "higher than the beasts, lower than the angels." Faced with the twin dangers of death and dehumanization, it will be important to advance an account of human dignity that does justice both to (a) the equal dignity we all share by virtue of our common humanity and (b) the dignity to which we can all aspire by exercising our humanity to the greatest and finest extent possible. Neither the dignity of equal humanity nor the dignity of human excellence is imageable in the Man of Glass.

There is no question that modern science is one of the truly great monuments to the human intellect. Precisely because it is value neutral and heuristically materialist, it gains the kind of knowledge of how our bodies work that is tremendously helpful in ameliorating disease and relieving suffering. But it cannot even come within hailing distance of human perfection, let alone salvation. In seeking our salvation, if salvation is to be sought, we must continue to look beyond ourselves and honor the longings of our souls, while humbly using our limited powers and still more limited wisdom to try to make our world a little bit better rather than a little bit worse. In the end, the good that we will do with science and medicine can only be completed by avoiding those evils that come from seeing health as salvation, the soul as biochemicals, and medicine as the messiah.

NOTES

1. A traveling version of the exhibit, first presented at the United States Holocaust Memorial Museum (USHMM) in Washington, D.C., has been seen also in Pittsburgh, Atlanta, and Minneapolis. A precise German translation of the exhibit opened in 2006 in Dresden. The USHMM website has a version of the exhibit at http://www.ushmm.org/museum/exhibit/online/deadlymedicine/. An earlier, shorter version of this chapter was presented at the USHMM in the "Insight" series connected with the exhibit on March 17, 2005.
2. René Descartes, "Discourse on the Method of Conducting One's Reason Well and Seeking Truth in the Sciences," trans. Richard Kennington, ed. Pamela Kraus and Frank Hunt. *Descartes: Discourse on Method* (Newburyport, MA: Focus Philosophical Library, 2007), Part VI, paragraph 2.
3. Karl Binding and Albert Höche, "Permitting the Destruction of Unworthy Life: Its Extent and Form," Essay One, trans. Walter E. Wright. *Issues in Law & Medicine*, 8, no. 2 (1992): 246. Originally published in German as "*Lebensunwerten lebens*" by Verlag von Felix Meiner. Leipzig, Germany (1920).
4. Karl Binding and Albert Höche, "Permitting the Destruction of Unworthy Life: Its Extent and Form," Essay Two, 265.
5. Matt. 5:48.
6. Robert L. Sinsheimer, "The Prospect of Designed Genetic Change," *Engineering and Science Magazine* (California Institute of Technology), April 1969. Dr. Sinsheimer subsequently had a change of heart, and he has become one of the advocates for caution and sobriety.
7. Joshua Lederberg, "Unpredictable Variety Still Rules Human Reproduction," *The Washington Post*, September 30, 1967.

8. Bentley Glass, "Science: Endless Horizons or Golden Age?" *Science* 171 (1971): 28.
9. Peter Singer, *Practical Ethics*, 2nd ed. (New York: Cambridge University Press, 1993), 186.
10. Marvin Olasky, "Blue-State Philosophy: Same-sex marriage? Euthanasia? Child's play issues in the avant-garde philosophy of Peter Singer," *World Magazine*, November 27, 2004, http://www.worldmag.com/articles/9987.
11. President's Council on Bioethics, *Beyond Therapy: Biotechnology and the Pursuit of Happiness*, October 2003. Available online at http://www.bioethics.gov/reports/beyond therapy, or commercially in trade editions, edited by Leon Kass, published by Dana Press and by Harper Collins.
12. International Academy of Humanism, "Statement in Defense of Cloning and the Integrity of Scientific Research," May 16, 1997, http://www.secularhumanism.org/library/fi/cloning_declaration_17_3.html.
13. Steven Pinker, "A Matter of Soul," Correspondence Section, *The Weekly Standard*, February 2, 1998, 6.
14. Prof. Pinker has responded to this critique (repeated in another essay of mine, "Science, Religion, and the Human Future," *Commentary*, April 2007) in the Letters section of the July/August 2007 issue of *Commentary*, which also contains, in response, my elaboration of this critique.

CHAPTER 11

WHAT DOES "MEDICINE AFTER THE HOLOCAUST" HAVE TO DO WITH AID IN DYING?

KATHRYN L. TUCKER

What place does a chapter discussing aid in dying, the choice made by a mentally competent, terminally ill patient to self-administer medication for the purpose of bringing about a peaceful death,[1] have in a collection of chapters on "Medicine after the Holocaust"? Most of us will appreciate immediately that the Nazi regime's program to terminate lives deemed unworthy by the state, including under the guise of "medical treatment," bears no relation to the choice of a terminally ill individual to choose to exercise a measure of control over his or her own impending death. This chapter seeks to illuminate the differences between these scenarios and to provide an overview of the background and status of aid in dying in the United States.

The need to distinguish these scenarios arose in a case before the U.S. Supreme Court known as *Glucksberg v. Washington* in the 1990s. In that case, a number of mentally competent, terminally ill patients asserted a right protected by the U.S. Constitution's guarantees of privacy and liberty to exercise a measure of control over the time and manner of their death.[2] Specifically, the patients claimed that to be able to obtain peaceful deaths they needed a prescription for life-ending medication, which could only be provided by a licensed physician. Choosing how to die when confronted by death due to terminal illness, the plaintiffs claimed, was a deeply personal decision, and should be reserved to the individual and not controlled by the state.

Dozens of briefs were filed with the Court opposing and supporting this position. Some of those opposed to the right claimed by the patients suggested that to permit this choice would be tantamount to Nazi-type eugenics. Noted Nazi historian and disability rights activist Hugh Gregory Gallagher[3] squarely rejected this argument in an *amicus curiae* (friend of the Court) brief filed with the Court, supporting recognition of the right to aid in dying. Professor Gallagher's personal statement is set forth in full:

> I am a writer and a historian. I am also a polio quadriplegic as a result of an attack of polio in 1952. For more than a decade, I have been studying and writing extensively on disability rights issues. I am particularly interested in the treatment that people with disabilities receive from the medical system. I am considered one of the foremost experts on the program authorized by Nazi Germany to "euthanize" over 200,000 of its citizens with disabilities. The Nazi's euthanasia program offers a horrible example of how easy it is to go wrong when the state or a group authorized by the state is allowed to assume the power to judge the worth of another.
>
> Ironically, this program is now being used by some as a justification to deny Americans in the terminal stage of illness the right to die with assistance. In fact, the German experience shows how important it is that the autonomy of people with disabilities be honored in all aspects of their lives. I do not believe that people with terminal illnesses should be denied the option of ending their lives in order to obtain relief from intolerable pain and suffering. This most personal of all decisions should rest between the person and his God.
>
> In my book, *By Trust Betrayed: Patients, Physicians, and the License to Kill in the Third Reich,* I describe in detail how the medical establishment in Germany, at its request, was authorized by the government to provide a "mercy" death for patients who in the judgment of their physicians had "lives not worth living." The program was authorized by Hitler in 1939 and placed under the direction of his personal physician. Although the program was called "euthanasia," the vast majority of the people killed were not terminally ill or in great pain or even requesting death. The killings began slowly, but soon enough, entire wards were being provided "final medical treatment." The patients were given no choice over whether to live or die.
>
> The case of assisted suicide[4] is quite different: the patient with a terminal illness retains complete choice over whether to live or to die. Neither the state nor the physician may decide, based on their conceptions of the individual's quality of life; the individual must assess his or her own quality of life. This is true whether or not the individual has a disability.
>
> In our generation, great strides have been made to welcome people with disabilities into society as equal members with equal rights. For the first time, people with disabilities have assumed control over their own lives, without fear of sterilization, internment, segregation and ostracism, and the denial to their rights to vote, hold property, enter legal contracts, and obtain public education, transportation and accommodation. Now they should no more be denied the right to obtain assistance in dying if they become terminally ill than should anyone else.
>
> To my mind, the issue comes down to control—control over one's Self. This control over Self is the very heart of the disability rights struggle. In Nazi Germany 60 years ago, people with disabilities were deprived of all control over their Selves. They were killed not because they sought death but because they did not measure up to "quality of life" standards set by their physicians with the concurrence of the state. This must never happen here.
>
> In the United States today, we are debating whether an individual in the terminal stages of an illness should retain control over his/her Self and his/her personal concept of quality of life—even to the point of death with dignity—or whether such control should be circumscribed by the state.

> In my own experience, it was this sense of control and the concomitant knowledge that I could end my life should the situation become unbearable, that kept me going through the extraordinary pain and suffering that accompanies acute polio. I will not turn over this control, or my personal decision to live or die, to the state. That decision is mine and mine alone.[5]

The *amicus curiae* brief joined by Professor Gallagher pointed out the several serious problems with analogizing the right sought by the terminally ill patients in *Glucksberg* to the Nazi eugenics program.

First, the point was made that the United States "is a nation dedicated to individual freedom in which the state is constitutionally limited in depriving individuals or life, liberty, or property and in which the press is constitutionally empowered to inform the citizenry of abuses of state power. The notion that the state will have an interest in killing, or authorizing the killing of, people with disabilities is ludicrous."[6]

Next, we were reminded of the historical fact that

> the Nazi program did not begin by granting a right to assisted suicide [*sic*] for people with terminal illnesses. It began with a determination by the state that people with disabilities are inferior, have an inferior quality of life, and are therefore disposable. That determination was in fact entirely consistent with the goals and ideals of the Nazi party, and flourished when the party flourished. By contrast, the notion that people with disabilities may be killed without their consent is, and will always be, abhorrent to the American public. The false notion that they are inferior or necessarily have an inferior quality of life may never in this country serve as a basis for any public policy, particularly a policy that would deprive them of life, liberty or property. The right being sought here is not in any way based on any perceived inferiority, or inferior quality of life of, people with terminal illnesses.[7]

Finally, it was noted that "the state in Nazi Germany gave the medical profession unbridled power and authority to control the lives and deaths of individuals with disabilities. In current cases, Respondents and *Amici* seek to expand individual autonomy and liberty, and to diminish the power and authority of the state to make or authorize decisions concerning the life and death of an individual. The implication of recognition of a right to assisted suicide is, therefore, not that the state will have more power to take the life of any of its citizens, with or without disabilities, but rather that it will have less power."[8]

In addition to these distinctions, it is notable that the Nazi euthanasia program was conducted in great secrecy, with every effort made to hide the program from the public. This cloak of secrecy marks another stark difference between the Nazi program and aid in dying as sought in the *Glucksberg* case and as practiced in the state of Oregon.

The U.S. Supreme Court resolved the *Glucksberg* case by inviting the states to grapple with the issue:

> Throughout the Nation, Americans are engaged in an earnest and profound debate about the morality, legality, and practicality of physician-assisted suicide. Our holding permits this debate to continue, as it should in a democratic society.[9]

The Court contemplated that on this controversial issue, one courageous state could address the issue, and the other states could watch and learn. This is exactly what has happened in Oregon. It is timely now, more than a decade after *Glucksberg,* to assess the lessons that have been learned from the experience in Oregon.

The Oregon Death with Dignity Act (Dignity Act), which began to be implemented in 1998, establishes tightly controlled procedures under which competent, terminally ill adults under the care of an attending physician may obtain a prescription for medication to allow them to control the time, place, and manner of their own impending death.[10] The attending physician must, among other things, determine that the patient is mentally competent and an Oregon resident, and confirm the patient's diagnosis and prognosis.[11] To qualify as "terminally ill," a person must have "an incurable and irreversible disease that has been medically confirmed and will, within reasonable medical judgment, produce death within six months."[12]

The attending physician must also inform any patient who requests such medication of their diagnosis and prognosis, the risks and probable results of taking the medication, and of alternatives, including, but not limited to, hospice care and pain relief.[13] A consulting physician must confirm the attending physician's medical opinion.[14]

If a qualifying patient makes a request under the Dignity Act, and it has been properly documented and witnessed and all waiting periods have expired,[15] the attending physician may prescribe, but not administer, medication to enable the patient to end his or her life in a humane and dignified manner. The Dignity Act immunizes physicians and pharmacists who act in compliance with its comprehensive procedures from civil or criminal sanctions and any professional disciplinary actions based on that conduct.[16] The Dignity Act requires health care providers to file reports with the state documenting their actions.[17]

This entire process, with many safeguards built in to ensure that the patient understands his or her medical condition and all options, is thus designed to empower the patient with information and a broad range of options, and lodges all decision-making power with the patient. The program is also designed to ensure that implementation is open to review and scrutiny. This openness and transparency, not to mention the locus of control, contrasts starkly with both the secrecy of the Nazi eugenics program and the vesting of decision-making power in the state, and its agents, to make life-ending decisions for individuals.

Oregon's experience with aid in dying has been extensively documented and studied. Each year the Oregon Health Division and Department of Human Services issues an annual report that presents and evaluates the state's experience with the Dignity Act.[18] Related reports and articles have also been published in leading medical journals.[19] These reports constitute the only source of reliable data regarding actual experience with legal, regulated physician-assisted dying in America.

The experience in Oregon has demonstrated that a carefully drafted aid-in-dying law does not place patients at risk.[20] In a report examining the Oregon experience to assess whether vulnerable populations were put at risk, the researchers concluded that there was no evidence of this.[21] The Oregon experience has caused even staunch opponents to admit that continued opposition to such a law can only be based on personal moral or religious grounds.[22]

The Oregon reports have shown the dire predictions of those initially opposed to the Dignity Act to have been unfounded. The data demonstrate that the option of physician-assisted dying has not been forced upon those who are poor, uneducated, uninsured, or otherwise disadvantaged.[23]

In fact, the studies show just the opposite. For example, the *Eighth Annual Report* found that a higher level of education is strongly associated with the use of physician-assisted dying; those with a baccalaureate degree or higher were 7.9 times more likely than those without a high school diploma to choose physician-assisted dying.[24] The report found that 100 percent of patients opting for physician-assisted dying under the Dignity Act had either private health insurance, Medicare, or Medicaid, and 92 percent were enrolled in hospice care.[25] Furthermore, the reports demonstrate that use of physician-assisted dying is limited. During the first nine years in which physician-assisted dying was a legal option, only 292 Oregonians chose it.[26] And, although there has been a gradual increase in the rate of those opting for physician-assisted dying, the overall rate remains low: the 38 terminally ill adults who chose this option in 2005 represented only 12 deaths for every 10,000 Oregonians who died that year.[27] A 2000 survey of Oregon physicians found that they granted one in six requests for aid in dying and that only one in ten requests resulted in hastened death.[28] Roughly one-third of those patients who complete the process of seeking medications under the Dignity Act do not go on to consume the medications.[29] These individuals derive comfort from having the option to control time of death, yet, ultimately, die of their disease without exercising that control.

In a recent study reviewing all annual reports of the Oregon Department of Human Services on utilization of the Oregon Death with Dignity Act, as well as in three independent studies, rates of assisted dying in Oregon were reviewed and showed no evidence of heightened risk for the elderly, women, the uninsured, people with low educational status, the poor, the physically disabled and chronically ill, minors, persons with psychiatric illnesses including depression, or racial minorities, compared to background populations. The authors concluded, "There is **no current evidence** where assisted dying is already legal for the slippery slope claim that legalized physician-assisted dying will have disproportionate impact on patients in vulnerable groups."[30]

Outside observers, after carefully studying implementation of the aid-in-dying law in Oregon, have concluded that the law poses no risk to patients. For example, the state of Vermont, after thoroughly reviewing the Oregon experience, concluded that "it is [quite] apparent from credible sources in and out of Oregon that the Death with Dignity Act has not had an adverse impact on end-of-life care and, in all probability, has enhanced the other options."[31] Leading scholars have reached similar conclusions: "I was worried about people being pressured to do this. But this data confirms . . . that the policy in Oregon is working. There is no evidence of abuse or coercion or misuse of the policy."[32]

Indeed, rather than posing a risk to patients or the medical profession, the Dignity Act has galvanized significant improvements in the care of the dying in Oregon. Oregon doctors report that since the passage of the Dignity Act, efforts have been made to improve their ability to provide adequate end-of-life care. These efforts include improving their knowledge of the use of pain medications for the terminally ill, enhancing their ability to recognize depression and other psychiatric disorders, and more frequently referring their patients to hospice programs.[33] One survey of Oregon physicians on their efforts to improve end-of-life care since 1994 found that 30 percent of respondents increased their number of referrals to hospice care, and 76 percent made efforts to increase their knowledge of pain medication.[34] A survey of hospice nurses and social workers in Oregon reveals that they observed an increase in physician knowledge of palliative care and willingness to refer and care for hospice patients from 1998 to 2003.[35]

In addition to the improvement of end-of-life care, the legal option of physician-assisted death has psychological benefits for both the terminally ill and the healthy. The availability of the option of aid in dying gives the terminally ill autonomy, control, and choice, which physicians in Oregon have identified as the overwhelming motivational factor behind the decision to request assistance in dying.[36] Healthy Oregonians know that if they ever face a terminal illness, they, too, can have control and choice over their manner of death.

The data demonstrate that, far from posing any hazard to patients or the practice of medicine, making the option of assisted dying available has galvanized improvements in end-of-life care and benefited all terminally ill Oregonians. A central argument against allowing patients access to aid in dying has been that risks would arise if the option were available.[37] Actual experience demonstrates that risk does not, in fact, exist. And the lack of risk undermines the argument against aid in dying.[38] This has led some major medical organizations to conclude that passage of Oregon-type aid-in-dying laws is good policy and to adopt policy supporting passage of such laws. For example, the American Medical Women's Association "supports the right of terminally ill patients to hasten what might otherwise be a protracted, undignified, or extremely painful death" and supports passage of Oregon-style aid-in-dying laws in other states.[39] Other major national medical and health policy organizations adopting policy in support of aid in dying include the American Medical Students' Association,[40] the American College of Legal Medicine,[41] and the American Public Health Association.[42] Others, recognizing the split in views on the issue among members, have adopted a neutral position, such as that taken by the American Academy of Hospice and Palliative Medicine.[43]

A fraction of dying patients, even with excellent pain and symptom management, confront a dying process so prolonged and marked by such extreme suffering and deterioration that they determine that hastening impending death is the least bad alternative. Passage of aid-in-dying laws harms no one and benefits both the relatively few patients *in extremis* who would make use of the option and a great many more who would draw comfort from knowing this is available should their dying process become intolerable to them.

Any one considering the issue now does so with more than a decade of data from the state of Oregon, which firmly puts to rest the concern that a legal option of aid in dying poses risk to patients or physicians. The question, finally, is simply, are we sufficiently compassionate to allow the choice of aid in dying to terminally ill, competent patients who are receiving state-of-the-art, end-of-life care but are still suffering? Other states are following Oregon's lead and making aid in dying legal.

Voters in the state of Washington considered the issue in 2008 and, after an intense campaign with opponents making the same claims about harm to patients and the medical profession, voted to adopt the Washington Death with Dignity Act[44] by the significant margin of 58 percent to 42 percent.[45] The Washington measure is virtually identical to Oregon's and began implementation in March 2009.

The experience in Washington with aid in dying will, no doubt, be closely watched and will contribute to the body of data on how the availability of this intervention impacts end-of-life care. Such data will then inform consideration of policy and legislation in other states.

As a result of a court case, Montana recently recognized that the freedom of its terminally ill citizens to choose aid in dying is a fundamental right protected by its state constitution's guarantees of privacy and dignity.[46]

What does "Medicine after the Holocaust" have to do with "Aid in Dying"? Nothing. Indeed, a critical lesson from the Holocaust is the paramount importance of protecting the right of individuals to control their own bodies and to never permit the state to determine when a life is, or is not, deemed sufficiently worthy. As the renowned legal philosopher Ronald Dworkin observes, and cautions, "Making someone die in a way that others approve but he believes a horrifying contradiction of his life, is a devastating, odious form of tyranny."[47]

NOTES

1. This choice has been referred to as "assisted suicide" in the past. That term has over time been recognized as inaccurate and emotionally charged by a large and growing number of major medical and health policy organizations. See, e.g. American Medical Women's Association, AMWA Position on Aid in Dying, http://www.amwa-doc.org/index.cfm?objectId=242FFEF5-D567-0B25-585DC5662AB71DF9; American Academy of Hospice and Palliative Medicine, AAHPM Position Statements on Physician-Assisted Death, approved February 2007, http://www.aahpm.org/positions/suicide.html (rejecting the term Physician-Assisted Suicide as "emotionally charged" and inaccurate); American Medical Student Association, Principles Regarding Physician Aid In Dying, http://www.amsa.org/about/ppp/pas.cfm; American College of Legal Medicine, ACLM Policy on Aid in Dying, http://www.aclm.org/resources/articles/ACLM%20Aid%20in%20Dying%20Policy.pdf; American Public Health Association, Patients' Right to Self-Determination at the End of Life, http://www.apha.org/advocacy/policy/policysearch/default.htm?id=1372.
2. *Glucksberg v. Washington,* 521 U.S. 702, 735 (1997).
3. Professor Hugh Gregory Gallagher is the author of *By Trust Betrayed* (Arlington, VA: Vandamere Press, 1995), a study of the Nazi eugenics and euthanasia program based on extensive research. Professor Gallagher served as legislative assistant to Senator Bob Bartlett of Alaska and was involved in passage of the Architectural Barriers Act, the first major federal legislation ensuring accessibility for persons with disabilities to public buildings and spaces.
4. See note 1 above regarding terminology. In the mid-1990s the term "assisted suicide" was most often used to refer to the practice that is now more commonly referred to as aid in dying. Those opposed to this option, including Wesley Smith, who is the author of a chapter in this book, continue to use the term "assisted suicide," recognizing that it is intentionally divisive. For an extensive discussion of the power of labeling in this context

see Kathryn L. Tucker, "Patient Choice at the End of Life: Getting the Language Right," *Journal of Legal Medicine* 28 (2007): 305–325.

5. Brief for Gay Men's Health Crisis and Lambda Legal Defense and Education Fund as Amici Curiae Supporting Respondents at Appendix, *Glucksberg v. Washington,* 521 U.S. 702 (1997) (No. 96–110), 1996 WL 711205.
6. Brief for Amici Curiae at Appendix, *Glucksberg,* 521 U.S. 702 (No. 96–110).
7. Ibid.
8. Ibid.
9. *Glucksberg v. Washington,* 521 U.S. 702, 735 (1997).
10. Or. Rev. Stat. §127.805 (2006).
11. Or. Rev. Stat. §127.815 (2006).
12. Or. Rev. Stat. §127.800(12) (2006).
13. Or. Rev. Stat. §127.800(7) (2006).
14. Or. Rev. Stat. §127.800(8) (2006); Or. Rev. Stat. §127.820 (2006).
15. Or. Rev. Stat. §§127.840–127.850 (2006). The Dignity Act requires a 15-day waiting period between the patient's initial oral request and the writing of the prescription, and a 48-hour waiting period between the patient's written request and the writing of the prescription. Or. Rev. Stat. §127.850 (2006).
16. Or. Rev. Stat. §127.885 (2006).
17. Or. Rev. Stat. §127.865 (2006).
18. See Oregon Department of Human Services, Death with Dignity Act: Records and reports data on the act, http://oregon.gov/DHS/ph/pas/index.shtml.
19. See, e.g. Margaret P. Battin et al., "Legal Physician-Assisted Dying in Oregon and the Netherlands: Evidence Concerning the Impact on Patients in 'Vulnerable' Groups," *Journal of Medical Ethics* 33 (2007): 591–597.

    This report also examined the experience in the Netherlands, and, while that program is quite different from Oregon's, the researchers found no evidence of harm to persons in vulnerable populations in the Netherlands either. See also Amy D. Sullivan, Katrina Hedberg, and David W. Fleming, "Legalized Physician-Assisted Suicide in Oregon—The Second Year," *New England Journal of Medicine* 342 (2000): 598; Arthur E. Chin et al., "Legalized Physician-Assisted Suicide in Oregon, The First Year's Experience," *New England Journal of Medicine* 340 (1999): 577; Andrew I. Batavia, "So Far So Good: Observations on the First Year of Oregon's Death with Dignity Act," *Psychology, Public Policy and Law* 6 (2000): 291; David Orentlicher, "The Implementation of Oregon's Death with Dignity Act: Reassuring, but More Data Are Needed," *Psychology, Public Policy and Law* 6 (2000): 489 (implementation of Oregon law has so far been limited to terminally ill patients with a clear, persistent, and voluntary request for hastened death); Timothy E. Quill and Christine K. Cassel, "Professional Organizations' Position Statements on Physician-Assisted Suicide: A Case for Studied Neutrality," *Annals of Internal Medicine* 138, no.3 (2003): 208–211; Joseph Straton, "Physician Assistance with Dying: Reframing the Debate; Restricting Access," *Temple Political & Civil Rights Law Review* 15 (2006): 475.
20. See Battin et al., "Legal Physician-Assisted Dying," 591–597. See also Quill and Cassel, "Professional Organizations' Position Statements," 208–211; Linda Ganzini et al., "Oregon Physicians' Attitudes about and Experiences with End-of-Life Care since Passage of the Oregon Death with Dignity Act," *Journal of the American Medical Association* 285 (2001): 2363–2369; Melinda A. Lee and Susan W. Tolle, "Oregon's Assisted Suicide Vote: The Silver Lining," *Annals of Internal Medicine* 124 (1996): 267–269; Straton, "Physician Assistance with Dying," 475.
21. Battin et al., "Legal Physician-Assisted Dying," 591–597.

22. See Daniel Lee, "Physician-Assisted Suicide: A Conservative Critique of Intervention," *Hastings Center Report* 33 (2003): 17–19.
23. See, e.g. Arthur Eugene Chin et al., "Oregon's Death with Dignity Act: The First Year's Experience," *Department of Human Resources Oregon Health Division Center for Disease Prevention and Epidemiology Annual Report* (1999): 7, http://oregon.gov/DHS/ph/pas/docs/year1.pdf ("Patients who chose physician-assisted suicide were *not* disproportionately poor [as measured by Medicaid status], less educated, lacking in insurance coverage, or lacking in access to hospice care.") See also Battin et al., "Legal Physician-Assisted Dying," 591–597; Kant Patel, "Euthanasia and Physician-Assisted Suicide Policy in the Netherlands and Oregon: A Comparative Analysis," *Journal of Health and Social Policy*, 19, no.1 (2004): 37 (finding no empirical evidence of slippery slope in Oregon, but more potential for a slide in the Netherlands).
24. Richard Leman, ed., "Eighth Annual Report on Oregon's Death with Dignity Act," *Oregon Department of Human Services Office of Disease Prevention and Epidemiology* (March 9, 2006): 12, http://oregon.gov/DHS/ph/pas/docs/year8.pdf.
25. Leman, "Eighth Annual Report," 23.
26. Oregon Department of Human Services, "Summary of Oregon's Death with Dignity Act-2006," *Death with Dignity Act Annual Report* (2006): 1, http://oregon.gov/DHS/ph/pas/docs/year9.pdf. Some commentators have observed that legal medical interventions that will bring about death, such as removal of feeding tubes, are reluctantly taken, and have reasoned from this that if aid in dying were legal it would also be rare. See David Orentlicher and Christopher Callahan, "Feeding Tubes, Slippery Slopes, and Physician-Assisted Suicide," *Journal of Legal Medicine* 25 (2004): 389. The Oregon data support this contention.
27. Leman, "Eighth Annual Report," 5.
28. Linda Ganzini et al., "Physicians' Experiences with the Oregon Death with Dignity Act," *New England Journal of Medicine* 342 (2000): 557 (finding that the availability of palliative care led some, but not all, patients to change their mind about hastened death).
29. See, e.g. Oregon Department of Human Services, *Death with Dignity Act Annual Report* (2006): 1, fig. 1 (showing number of recipients of prescription each year, compared to number of deaths from use of prescription.).
30. Battin et al., "Legal Physician-Assisted Dying," 591–597.
31. Robin Lunge, Maria Royle, and Michael Slater, Oregon's Death with Dignity Law and Euthanasia in the Netherlands: Factual Disputes 2004, 30, http://www.leg.state.vt.us/reports/04Death/Death_With_Dignity_Report.htm.
32. William McCall, "Assisted-Suicide Cases Down in '04," *The Columbian*, sec. C, March 11, 2005, quoting Arthur Caplan, director of the Center for Bioethics at the University of Pennsylvania School of Medicine. See also Straton, "Physician Assistance with Dying," 475.
33. Straton, "Physician Assistance with Dying," 475. See also Quill and Cassel, "Professional Organizations' Position Statements," 208–211; Ganzini et al., "Oregon Physicians' Attitudes," 2363–2369; Lee and Tolle, "Oregon's Assisted Suicide Vote," 267–269; Lawrence J. Schneiderman, review of *Physician-Assisted Dying: The Case for Palliative Care and Patient Choice*, Timothy E. Quill and Margaret P. Battin, eds. *Journal of the American Medical Association*, 293 no. 4 (2005): 501 ("Indeed, one of the unexpected yet undeniable consequences of Oregon's Death with Dignity Act permitting physician aid in dying is that 'many important and measurable improvements in end-of-life care' occurred following the Act's implementation. Rather than becoming the brutal abattoir for hapless patients

that some critics predicted, the state is a leader in providing excellent and compassionate palliative care.")

34. Ganzini et al., "Oregon Physicians' Attitudes," 2363–2369.
35. Elizabeth R. Goy et al., "Oregon Hospice Nurses and Social Workers' Assessment of Physician Progress in Palliative Care Over the Past 5 Years," *Palliative and Supportive Care* 1 (2003): 215.
36. Kathy L. Cerminara and Alina Perez, "Empirical Research Relevant to the Law: Existing Findings and Future Directions, Therapeutic Death: A Look at Oregon's Law," *Psychology, Public Policy and Law* 6 (2000): 512–513 (acknowledging possible negative effects of legalized aid in dying, but the data from Oregon in one year justifies optimistic view). See also Linda Ganzini et al., "Oregon Physicians' Perceptions of Patients Who Request Assisted Suicide and Their Families," *Journal of Palliative Medicine* 6 (June 2003): 381 (finding physicians receiving requests for lethal medication perceive patients as wanting to control their deaths); Linda Ganzini et al., "Experiences of Oregon Nurses and Social Workers with Hospice Patients Who Requested Assistance with Suicide," *New England Journal of Medicine* 347 (2002): 582 (nurses and social workers rated desire to control circumstances of death as most important reason for requesting aid in dying).
37. See, e.g. *Washington v. Glucksberg,* 521 U.S. 702, 732 (1997) ("We have recognized . . . the real risk of subtle coercion and undue influence in end-of-life situations.")
38. See Kathryn L. Tucker, "The Chicken and the Egg: The Pursuit of Choice for a Humane Hastened-Death as a Catalyst for Improved End-of-Life Care; Improved End-of-Life Care as a Precondition for Legalization of Assisted Dying," *New York University Annual Survey of American Law* 60 (2004): 355. Other reasons that put to rest the fear that passage of aid-in-dying laws will put patients at risk have been offered. For example, one commentator studied the reluctance of patients and providers to withdraw feeding tubes, an option legal in every state. He concluded that the data show that feeding tubes are overutilized and from this argues that this demonstrates reluctance to take steps that will precipitate death, and that such reluctance will apply in context of aid in dying. See Orentlicher and Callahan, "Feeding Tubes," 389.
39. American Medical Women's Association, AMWA Position on Aid in Dying, approved September 9, 2007, http://www.amwa-doc.org/index.cfm?objectId=242FFEF5-D567-0B25-585DC5662AB71DF9 (supporting the passage of aid-in-dying laws, which empower mentally competent, terminally ill patients and protect participating physicians, such as the Oregon Death with Dignity Act.) See also "Physicians group backs choices bill," *Eureka Reporter,* March 22, 2007, http://www.deathwithdignity.org/news/news/eurekareporter032207.asp (noting support by the California Association of Physician Groups for passage of an Oregon-type aid-in-dying law in California).
40. American Medical Student Association, Principles Regarding Physician Aid in Dying, http://www.amsa.org/about/ppp/pas.cfm.
41. American College of Legal Medicine, ACLM Policy on Aid in Dying, http://www.aclm.org/resources/articles/ACLM%20Aid%20in%20Dying%20Policy.pdf.
42. American Public Health Association, Patients' Right to Self-Determination at the End of Life, http://www.apha.org/advocacy/policy/policysearch/default.htm?id=1372.
43. American Academy of Hospice and Palliative Medicine, AAHPM Position Statements on Physician-Assisted Death, approved February 2007, http://www.aahpm.org/positions/suicide.html.
44. Wash. Rev. Code §§ 70.245.010-.904 (2009).

45. Washington Secretary of State, Results of November 4, 2008, General Election: Initiative 1000 concerns allowing certain terminally ill competent adults to obtain lethal prescriptions, http://vote.wa.gov/Elections/WEI/Results.aspx?RaceTypeCode=M&JurisdictionTypeID=-2&ElectionID=26&ViewMode=Results.
46. *Baxter v. Montana,* No. ADV-2007-787 (Mont. 1st Jud. Dist., December 5, 2008).
47. Ronald Dworkin, *Life's Dominion: An Argument about Abortion, Euthanasia, and Individual Freedom* (New York: Alfred A. Knopf, 1993), 217.

CHAPTER 12

Is Physician-Assisted Suicide Ever Permissible?

WESLEY J. SMITH

As we travel ever deeper into the new century and millennium, the struggle over legalizing assisted suicide is increasingly taking center stage. The cultural stakes could not be higher. The manner in which the nation ultimately decides this crucial legal and moral issue will test our culture's commitment to the sanctity/equality of human life; define for our posterity society's cultural attitudes toward elderly, disabled, and dying people; and determine the continuing commitment of medicine to the "do no harm" ethical ideal embodied in the Hippocratic Oath.

Given the import of the question—whether assisted suicide should ever be "permissible"—the decision we make as a society must be based upon broad public policy considerations rather than worst-case-scenario, individual anecdotes. For example, on several occasions I have been challenged in public forums by assisted suicide supporters about a scene from the movie *The Sand Pebbles*. In the picture, the Steve McQueen character shoots and kills his Chinese friend who is being publicly tortured to death "by the death of a thousand cuts." The clear implication of the query seems to be that if the fictional shooting was right, so too could be euthanasia.

My answer to these challenges, and the premise of this chapter, is that such unique scenarios and even true anecdotes about individual agonizing deaths—which it must be conceded do happen—are not a proper basis for making crucial public policy decisions. In this regard, it is important to note that the statistics published in Oregon clearly indicate that people do not seek assisted suicide because of unrelievable agony. Rather, the suicidal patients primarily are worried about losing control, being unable to engage in enjoyable activities, or being burdens. These are important matters that require concentrated intervention by caregivers, but are not the kind of unbearable agony that proponents often cite as reasons for legalizing assisted suicide.[1] Rather, in determining our future course, we should focus primarily on how legalizing assisted suicide would affect the general welfare—with particular emphasis on how it could impact the care and well-being of the weakest and most vulnerable members of society. Happily, when it comes to euthanasia and assisted suicide, what is good for the whole also turns out to be the best protection for the lives and dignity of terminally ill, disabled, and otherwise suffering individuals.

Assisted Suicide, Not "Aid in Dying"

Before launching into the heart of this chapter, a few words about lexicon and nomenclature are in order. Historically, proponents of legalizing euthanasia and assisted suicide have manipulated words and redefined terms as a political tactic to gain public favor. Indeed, the word "euthanasia"—which is now synonymous with mercy killing—once meant dying naturally without pain surrounded by family. This switch in meanings was successfully accomplished with the stroke of the pen in 1870 when an English schoolteacher named Samuel Williams published an essay advocating mercy killing as a proper and acceptable act. As the historian Ian Dowbiggen has noted,

> In advocating voluntary active euthanasia and physician-assisted suicide, Williams was instrumental in redefining euthanasia as an act of mercy killing rather than a passive process in which the discomforts of death are mitigated but not intentionally ended by painkillers.[2]

This word engineering has continued ever since.[3] The latest iteration of this approach is the new euphemistic advocacy term "aid in dying." U.S. advocates for legalizing assisted suicide claim that since most current legalization proposals limit those eligible to receive lethal prescriptions to patients diagnosed with a terminal condition—that isn't necessarily true in other countries—the act of self-killing by this cohort should not be considered "suicide," but something else that doesn't carry with it the same negative connotations among the general public.[4]

In fact, advocates for legalizing assisted suicide object to the word "suicide" and the term "assisted suicide" precisely because these words are accurate and descriptive. Indeed, this very point was made by one of the patriarchs of the modern assisted suicide movement, the founder of the Hemlock Society, Derek Humphry, who wrote to the *Register Guard* about a scheme in Oregon to rename assisted suicide "death with dignity," words equally applicable to "aid in dying:"

> As the author of four books on the right to choose to die, including *Final Exit,* I find the vacillation by the Department of Human Services (Register-Guard, Oct. 23) on how to describe the lawful act of a physician helping a terminally ill person to die by handing them a lethal overdose, which they can choose to drink (or not), an affront to the English language.
>
> "Physician" means a licensed M.D.; "assisted" means helping; and "suicide" means deliberately ending life. The department's cop-out choice of the words "death with dignity" is wildly ambiguous and means anything you want. Let's stick to the English language and in this matter call a spade a spade.[5]

When a movement feels the need to hide what is being advocated behind a smoke screen of obfuscating language and gooey euphemisms, then it is reasonable to conclude that there is something wrong with what is being advocated. Thus, throughout this chapter the term "assisted suicide" will be used precisely because it is accurate, descriptive, and because in such an important public policy debate, as Humphry wrote, it is important to call a spade a spade.

The Consequences of Accepting Assisted Suicide Ideology

To understand the profound threat that legalizing assisted suicide poses to the weak and vulnerable, we need to explore the movement's ideological presumptions. Two fundamental ideological bases are asserted by advocates for legalizing assisted suicide. The first is a fervent embrace of radical individualism, which perceives the right of personal autonomy as being virtually absolute. Accordingly, promoters of assisted suicide generally believe that "the individual's right to self-determination—to control the time, place, and manner of death" is a paramount liberty interest, sometimes called "the ultimate civil right." The second ideological principle underlying assisted suicide advocacy is that killing (ending life) is an acceptable answer to the problem of human suffering.[6]

Once these ideological premises are accepted—that autonomy is paramount and killing is a valid answer to human suffering—restricting assisted suicide to the dying becomes utterly illogical. After all, many people experience far greater suffering and for a far longer period than other people who are terminally ill. Thus, once the premises of assisted suicide advocacy become accepted by a broad swath of the medical professions and the public, there is little chance eligibility for such "permitted" suicide will remain strictly limited. Indeed, even before assisted suicide has been accepted widely, the gravity-like pull toward broadening eligible categories for assisted suicide eligibility has already begun. For example, writing recently in the prestigious bioethics journal the *Hastings Center Report*, bioethicist Jacob M. Appel advocated expanding assisted suicide *to the mentally ill*:

> At the core of the argument supporting assisted suicide are the twin goals of maximizing individual autonomy and minimizing human suffering. Patients, advocates believe, should be able to control the decision of when to end their own lives, and they should be able to avoid unwanted distress, both physical and psychological.

Noting that the Swiss Supreme Court recently ruled that mentally ill people have a constitutional right to assisted suicide, Appel, arriving at the logical conclusion of assisted suicide ideology, claims that granting the mentally ill access to assisted suicide would "empower" such patients:

> Most likely, the taboo against assisted suicide for the mentally ill is a well-meaning yet misplaced response to the long history of mistreatment that those with psychiatric illness have endured in western societies. Psychiatrists and mental health advocates may fear that their patients will be coerced to "choose death" against their wishes, or that, once suicide is an acceptable option, the care for those who reject assisted suicide will

> be diminished. But as the plaintiff argued before the Swiss high court, in challenging "medical paternalism," we are entering an era during which psychiatric patients do not need to be protected, but empowered. Our goal should be to maximize the options available to the mentally ill.[7]

Appel's logic is impeccable. Once assisted suicide is transformed into a palliative "medical treatment" for one classification of suffering patients, what, other than crass political expediency, is there to prevent access to life-terminating "treatment" to be made available to other suffering people?

THE DUTCH STORM WARNING

We need only look to the experience of the Netherlands to see the destructive social forces that assisted suicide ideology unleashes. The Dutch have permitted euthanasia and assisted suicide since 1973. Euthanasia and assisted suicide became an integral part of Dutch medicine after a court ruling that refused to meaningfully punish a physician, Geetruida Postma, who had euthanized her mother.[8] Other court decisions soon followed, with each widening and further liberalizing the conditions under which euthanasia and assisted suicide would not be punished. Thus, even though these life-terminating practices remained technically illegal until 2002, they became deeply entrenched in Dutch medical practice.

At this point it is important to recall that when euthanasia was first accepted in the Netherlands, it was supposed to be a rare event, to be resorted to only in the most unusual cases of "intolerable suffering"—just as we are told it will be by assisted suicide advocates in the United States. To ensure against abuses, "strict guidelines" were instituted in law, just as domestic proponents advocate.

But the guidelines have proved virtually worthless. Over time—precisely *because* of the fervent belief in "choice" and the widespread acceptance of assisted suicide and mercy killing as acceptable answers to human suffering—doctors began to interpret these death regulations loosely and even ignore them altogether. In the few circumstances where the law took notice, the courts accommodated expanded euthanasia through continual loosening of the meaning of the guidelines. Indeed, a statistical analysis of Dutch euthanasia practices published in *Journal of Medical Ethics* concluded that the Dutch promise of "effective regulation ring hollow" and that euthanasia in the Netherlands "remains beyond effective control."[9]

That is a tactful way of putting it. Since 1973, Dutch doctors have gone from euthanizing or assisting the suicides of the terminally ill who ask for it, to the chronically ill who ask for it, to the disabled who ask for it, to depressed people not physically ill who ask for it.[10] Assisting the suicides of the deeply depressed who are not physically ill received the imprimatur of the Dutch Supreme Court in the death of Hilly Bosscher, a profoundly depressed woman who wanted to kill herself because she had lost her two sons—one to suicide in 1986 and the other to brain cancer in 1991. Obsessed with joining her children in death, Bosscher moved the graves of her two sons to the same cemetery and purchased a burial plot for herself so that she could be buried between them. Rather than treat her, her psychiatrist assisted her suicide with the eventual approval of the court.[11]

The Dutch experience shows that once intentionally causing death becomes a mere "medical treatment," even the issue of "choice" is eventually subsumed. In the Netherlands, infants are killed because they have birth defects, and doctors justify the practice. A 1997 study published in the British medical journal *The Lancet* revealed how deeply pediatric euthanasia had metastasized into Dutch neonatal medical practice. According to the report, doctors killed approximately 8 percent of all infants who died in the Netherlands in 1995. Assuming this to be typical, this amounts to approximately 80–90 infanticides per year. Of these, one-third would have lived more than a month. At least 10–15 of these killings involved infants who did not depend on life-sustaining treatment to stay alive. The study found that 45 percent of neonatologists and 31 percent of pediatricians, who responded to the study's questionnaires, had killed infants.[12] A follow-up study of end-of-life decisions made for infants published in the April 9, 2005, issue found that nothing had changed. In 2001, "in 8 percent" of cases, drugs were administered to infants "with the explicit intention to hasten death."[13]

In 2004 Groningen University Medical Center made international headlines when it admitted to permitting pediatric euthanasia and published the "Groningen Protocol," infanticide guidelines the hospital had utilized when killing 15–20 disabled newborns.[14] The protocol creates three categories of killable infants: infants "with no chance of survival," infants with a "poor prognosis and are dependent on intensive care," and "infants with a hopeless prognosis," including those "not depending on intensive medical treatment but for whom a very poor quality of life... is predicted."[15]

In addition to killing babies, Dutch doctors have long been known to routinely euthanize patients who have not asked to die. Indeed, as far back as 1991, the Remmelink Report (named after the former Dutch attorney general) found *thousands* of Dutch patients who never asked to be euthanized have been killed by doctors.[16] These findings are startling—and are a vivid demonstration of the very slippery slope in the Netherlands that physician-assisted suicide advocates deny exists. According to the Remmelink Report, about 130,000 people die each year in the Netherlands.[17] Of these, approximately 43,300, or about one-third, die suddenly—from catastrophic heart attacks, stroke, accidents, et cetera—thus precluding medical decision making about end-of-life care.[18] That leaves approximately 90,000 people whose deaths involve end-of-life medical decision making each year.

With that in mind, here are the figures about euthanasia-related deaths in 1990, derived from the Remmelink Report's published statistical data:

- Of 2,700 doctor-induced deaths, 2,300 patients were euthanized by their doctors upon request, and 400 people died through physician-assisted suicide.[19] Therefore, approximately 3 percent of all deaths involving end-of-life medical care in the Netherlands were doctor-induced deaths (equivalent to 41,500 deaths in the United States).
- Involuntary euthanasia, lethal injections given without request or consent, accounted for 1,040 deaths—three deaths every single day.[20] These deaths constitute slightly more than 1 percent of all cases involving end-of-life

medical care. (The same percentage in the United States would translate into approximately 16,000 involuntary killings per year.)

- Of these involuntary euthanasia cases, 14 percent, or 145 cases, were fully competent to make their own medical decisions but were killed without request or consent.[21] (The same percentage in the United States would represent more than 2,000 people killed.) Moreover, 72 percent of the people killed without their consent had never given any indication they would want their lives terminated.[22]
- An intentional overdose of morphine or other pain-control medications, designed primarily to terminate life, accounted for 8,100 deaths.[23] In other words, death was not a side effect of treatment to relieve pain, which can sometimes occur, but was the *intended result* of the overdose. Of these, 61 percent (4,941 patients) were intentionally overdosed without request or consent. (The equivalent percentage in the United States would convert into approximately 78,000 deaths.)

These figures are startling. Of the approximately 90,000 Dutch people whose deaths involved end-of-life medical decision making in 1990, 11,140 were intentionally killed (euthanized) or assisted in suicide—or 11.1 percent of all Dutch deaths involving medical decision making! This is approximately 8.5 percent of Dutch deaths from all causes. Of these killings, *more than half were involuntary* (1,040 involuntary lethal injections and 4,941 involuntary intentional overdoses). Applying those percentages to the U.S. death rate would mean more than 170,000 deaths each year caused by euthanasia or assisted suicide, and about 85,000 of these involuntary, more than the current number of U.S. suicides and homicides combined.[24] (Subsequent studies have indicated fewer terminations without request or consent, but it is indisputable that many thousands of Dutch patients have been euthanized since euthanasia commenced in the Netherlands—even though such killings are completely illegal.)

Further studies, which were not so detailed about intentional overdoses of pain medication, still found that each year about 900 Dutch patients are nonvoluntarily euthanized by their doctors. The practice even has a name, "termination without request or consent," and even though this is formally considered murder under Dutch law, it is rarely prosecuted and almost never meaningfully punished.

Dr. Herbert Hendin, a psychiatrist and former medical director of the American Foundation for Suicide Prevention, is one of the world's foremost experts on Dutch euthanasia. Over the last several years, Hendin has held extensive discussions with Dutch doctors who euthanize patients and has reviewed the records of actual cases. Hendin believes that many doctors in the Netherlands feel justified in performing nonvoluntary euthanasia, because a system that permits them to kill "encourages some to feel entitled to make [euthanasia] decisions without consulting the patient."[25] As an example, Hendin recounts his interview with a pro-euthanasia doctor who justified killing a nun who had requested not to be killed on the basis of religious belief, because *he felt* she was in too much pain.[26]

Making matters worse—and perhaps most germane to the question under discussion—despite the clear abuses in the Netherlands, despite the uncontestable fact that doctors are now euthanizing babies and people who have not asked to be killed, the Dutch people support their country's euthanasia policy. And here we see the greatest problem with opening the door to suicide as an answer to the problem of suffering caused by illness. Once the law states that assisted suicide is right, the people's own values may soon follow—opening the door to a fall off the vertical moral cliff.

The "Oregon Defense"

When faced with these undeniable facts from the Netherlands, American assisted suicide advocates pull out the "Oregon defense," in which they claim that there have been zero abuses with Oregon's assisted suicide practice, and thus, whatever has happened in the Netherlands (and in other assisted suicide friendly venues such as Switzerland and Belgium), Oregon proves that in America there is nothing to fear. Despite these blithe assurances by assisted suicide defenders, there is no way to know how the law is actually working in Oregon since almost all of the information the state garners about assisted suicide depends on self-reporting by lethally prescribing doctors—who are about as likely to tell the state that they violated the law as they are to tell the IRS they cheated on their taxes. Further demonstrating the empirical unreliability of the "Oregon statistics," state agencies have no budget or authority to engage in any meaningful oversight of assisted suicide activities. As Dr. Katrina Hedberg—the lead author of most of the Oregon official annual reports—admitted to a British House of Lords fact-finding committee, "Not only do we not have the resources to do it [independently investigate assisted suicide] but we do not have any legal authority to insert ourselves."[27] Moreover, even the information that the state does collect is *destroyed* once the annual reports are published—meaning there is no way to independently verify the data that Oregon has reported since its law went into effect in 1997.[28]

What little we know about actual assisted suicide cases in Oregon (as opposed to statistical data) comes from media reports (usually facilitated by assisted suicide proponents) and one peer-reviewed study of the actual medical records of a patient who received a lethal drug-overdose prescription. Even under the most generous interpretation, the actual practice of assisted suicide should make us far from sanguine in following the state's example.

To pass Measure 16 (the initiative to legalize assisted suicide), proponents assured wary voters that physician-assisted suicide would be limited to those rare cases in which patients were in "severe, unrelenting, and intolerable suffering" that could not be otherwise relieved. Moreover, hastened deaths were only supposed to take place "in the context of a meaningful doctor-patient relationship," and then only after a deep and thorough discussion of all options between trusting patients and devoted physicians.[29]

But from the very beginning that has not been the actual practice. Much is known about the first reported legal assisted suicide because the assisted suicide advocacy organization Compassion in Dying (now called Compassion and Choices) held a press conference shortly after her death to provide details. According to the report, "Mrs. A" had terminal breast cancer. She did not swallow physician-prescribed poison because of unbearable suffering and agony. Rather, in her own words played on audiotape posthumously at the press conference, she wanted to "be relieved of all the stress I have."[30] But stress caused by dying and growing debilitation, while certainly a very real and substantive medical issue that needs to be taken seriously by caregivers, is a treatable condition that does not require killing to alleviate.

A subsequent in-depth analysis of this case by medical and bioethics experts revealed an even more detailed account of these troubling events. Upon receiving her terminal diagnosis, Mrs. A asked her treating doctor to assist in her suicide. The doctor refused. She consulted with a second doctor who also declined and diagnosed her as depressed. She then contacted Compassion in Dying, whose medical director Dr. Peter Goodwin spoke with her twice on the telephone, after which he decided that she wasn't depressed but merely "frustrated." Goodwin then referred her to Dr. Peter Reagan, a doctor who Goodwin knew would be willing to prescribe lethally.

Reagan referred her to a psychiatrist, who saw her only once, and a second doctor to confirm the terminal diagnosis. He also conducted a "cursory" discussion with the patient about alternatives to assisted suicide. When she voiced fears of being kept alive by artificial nutrition if she did not kill herself, the death doctor failed to assure her that she had the right to refuse such care—perhaps a crucial factor in her decision to commit assisted suicide. The woman did not know her prescribing doctor well, and indeed, *died a mere two-and-a-half weeks after their first meeting*, at a time when she was not in pain and still looked after her own house.[31]

Then there was the very disturbing case of Kate Cheney, reported in the *Oregonian*, which provided a sickening glimpse of the porous nature of the protective guidelines.[32] Cheney, age 85, was diagnosed with terminal cancer and sought assisted suicide. But there was a problem: Cheney was in the early stages of dementia, raising significant questions about her mental competence. So, rather than prescribe lethal drugs, her doctor referred her to a psychiatrist.

Her daughter, Ericka Goldstein, accompanied Cheney to the psychiatric consultation. The psychiatrist found that Cheney had a loss of short-term memory causing the psychiatrist to write in her report that while the assisted suicide seemed consistent with Cheney's values, "she does not seem to be explicitly pushing for this." The psychiatrist also determined that she did not have the "very high capacity required to weigh options about assisted suicide." Worse, the person who seemed most intent on Cheney committing assisted suicide wasn't the elderly patient but her daughter. Accordingly, the psychiatrist nixed the lethal prescription.

Advocates of legalized assisted suicide might, at this point, smile happily and point out that such refusals are the way the law is supposed to operate to protect the vulnerable. But that wasn't the end of Kate Cheney's story. According to the *Oregonian* report, Cheney appeared to accept the psychiatrist's verdict, but her daughter most explicitly did not. To circumvent the rejection of assisted suicide, Cheney's daughter merely did what anyone in Oregon wanting assisted suicide can do if refused by one physician: she went doctor shopping.

Kaiser Permanente, Cheney's HMO, acceded to Goldstein's demand for another opinion on her mother's mental competency. This time, the psychiatric consultation was with a clinical psychologist rather than with an M.D. psychiatrist. Like the first psychiatrist, this consulting psychologist found that Cheney had significant memory problems. For example, she could not recall when she had been diagnosed with terminal cancer. The psychologist also worried about familial pressure, writing that Cheney's decision to die "may be influenced by her family's wishes." Still, despite these reservations, the psychologist determined that Cheney was competent to commit suicide.

The final decision to approve the death was made by a Kaiser HMO ethicist/administrator named Robert Richardson. Dr. Richardson interviewed Cheney, who told him she wanted the poison pills, not because she was in irremediable pain, but because she feared not being able to attend to her personal hygiene. After the interview, satisfied that she was competent, he approved the lethal prescription.

It is important to reiterate that whatever protection the guidelines provide in Oregon ends with the writing of the lethal prescription. At that point, no doctor was required to be at the patient's bedside. Indeed, the law does not require that anything be done thereafter to determine if the patient is competent when swallowing the drug overdose or to prevent that patient from being coerced into taking the pills. In short, once Kate Cheney received the prescription, under the law, she was on her own.

What happened next in the Cheney case illustrates the potential for problems that can arise after the lethal prescription is issued. Cheney did not take her lethal drugs right away. According to the *Oregonian* report—which it must be recalled was based on the family's description—she first asked to die immediately after her daughter had to help her shower after an accident with her colostomy bag. But she quickly changed her mind. Then, Cheney was sent to a nursing home for a week so that her family could have some respite from caregiving.

The time spent in the nursing home may have pushed Cheney into wanting immediate death. As soon as she was brought home, she declared her desire to take the pills. Her grandchildren were quickly called to say their goodbyes, and Cheney swallowed her prescribed poison. She died with her daughter at her side, telling her what a courageous woman she was.

If Cheney was depressed when she swallowed the poison, there was no doctor available to diagnose it. If she was coaxed or pressured into taking the pills (which was not contended or implied in the *Oregonian* story), there were no witnesses from outside the family to protest. Indeed, other than what family members told the *Oregonian* reporter, we don't know what happened at Kate Cheney's death since the Oregon guidelines do not require any independent assessment of assisted suicide deaths.

Perhaps even more disturbing is the case of Michael Freeland, who appears to have been permitted to keep his lethal overdose by his psychiatrist after becoming psychotic. In contemplating the Oregon experience, it is worth nothing that Freeland's story was the first and, to the best of my knowledge, remains the only case in which the patient's medical records were available for professional perusal. The researchers found—*and this was never reported in the official Oregon statistics*—that not only was Freeland apparently not terminally ill as defined by Oregon's law (which requires that a doctor reasonably believe that a patient will die within six months) when he first received his lethal prescription, but even more alarmingly, he was allowed to keep his cache of suicide pills even after being diagnosed as having "depressive disorder," "chronic adjustment disorder with depressed mood," and "intermittent delirium," for which he was hospitalized and declared mentally incompetent by a court.[33]

Here are the details: Freeland was diagnosed with lung cancer in 2000. In early 2001, he received a lethal prescription from a physician referred by Compassion in Dying. On January 23, 2002, more than a year after receiving the lethal prescription, Freeland was admitted to a hospital psychiatric ward for depression with suicidal and possibly homicidal thoughts. A social worker went to Freeland's home and found it "uninhabitable," with "heaps of clutter, rodent feces, ashes extending two feet from the fireplace into the living room, lack of food and heat, et cetera. Thirty-two firearms and thousands of rounds of ammunition were removed by the police." Amazingly, the "lethal medications" that had been prescribed more than a year before were left in the house.

Freeland was hospitalized for a week and then discharged on January 30. The discharging psychiatrist noted with approval that his guns had been removed, "which resolves the major safety issue," but wrote that Freeland's lethal prescription remained "safely at home." Freeland was permitted to keep the overdose even though the psychiatrist reported he would "remain vulnerable to periods of delirium." In-home care was considered likely to assist with this problem, but a January 24 chart notation noted that Freeman "does have his life-ending medications that he states he may or may not use, so that [in-home care] may or may not be a moot point."[34]

The day after his discharge, the psychiatrist wrote a letter to the court in support of establishing a guardianship for Freeland, stating, "He is susceptible to periods of confusion and impaired judgment." The psychiatrist concluded that Freeland was unable to handle his own finances and that his cognitive impairments were unlikely to improve. He lived under supervision for a brief time, but was soon home alone with ready access to his suicide drugs.

It was during this period that Freeland called Physicians for Compassionate Care for help. Rather than dying alone by assisted suicide, he was instead cared for by psychiatrist N. Gregory Hamilton; Hamilton's wife, Catherine; and by his friends—who assured the now imminently dying man "that they valued him and did not want him to kill himself." Freeland was properly treated for depression with medication. He received good pain control, including a morphine pump. Best of all, he was reunited with his estranged daughter and died knowing she loved him and would cherish his memory.[35]

Based on their review of the facts and circumstances surrounding Freeland receiving a lethal prescription and being allowed to keep the drugs while psychotic, the Hamiltons reached important conclusions about the law's discriminatory effect on patients and its impact on mental health professionals:

> The legalization of doctor-assisted suicide in Oregon has resulted in the introduction of competing paradigms—the traditional clinical approach [removing lethal means is central to the clinical treatment of suicide symptoms] and the assisted-suicide competency model [providing lethal means]—for responding to suicidal thoughts and behaviors in seriously ill individuals... These competing models appear to be based on incompatible underlying assumptions about the value of protecting life depending on predictions of how long a patient might live... We conclude that the attempt to mix

> models is confusing to both clinicians and patients and endangers seriously ill patients, particularly those with a history of pre-existing mental illness.[36]

Or to put it another way, the very mental health professionals responsible for treating this delusional man expressed utter indifference to his committing suicide, whereas had he presented the same symptoms without having cancer, he would undoubtedly have received rigorous suicide prevention and treatment. The only conclusion to draw from such a disparity is that because he had obtained a suicide prescription, the value of Mr. Freeland's life was perceived to be of lesser value than other people by his own psychiatrist.

A DYSFUNCTIONAL HEALTH CARE SYSTEM

Beyond the ideology of the assisted suicide movement and the statistics and individual cases of assisted suicide in Oregon and the Netherlands, we must also ponder the impact of transforming assisted suicide from a crime (in almost all states) into a legitimate "medical treatment" in the context of America's dysfunctional health care system. Arguments in favor of assisted suicide almost always depict the act as occurring in an idealized world that *does not exist*. We are told to presume that decisions to commit assisted suicide would take place in the bosoms of loving families, that suicides would be facilitated by family doctors who have known the patient for decades and are intimately familiar with their values, and that assisted suicide would only be used as a last resort engaged in with great reluctance when nothing else could be done to alleviate unbearable pain.

But this assumes a world that doesn't exist. There are nearly 50 million people without health insurance in the United States—which means almost by definition that they may not have access to quality medical treatment and proper care. The economics of health care are increasingly driven by the HMO model in which profits are made by cutting costs: a system that many have complained bitterly has resulted in the chipping away of quality care. Some doctors are so stressed by the current system that patients may have a mere 15 minutes or less within which to interact with their doctors.

The cost for the drugs used in assisted suicide is less than $100. It could take $100,000 to provide the patient with proper care so they don't want assisted suicide. Should assisted suicide become legalized and legitimized, the economic force of gravity is obvious. After all, what could be a "cheaper" medical treatment than hastened death?

This point has even been acknowledged by zealous assisted suicide advocates like Derek Humphry, the founder of the Hemlock Society (now merged into Compassion and Choices), who wrote in his most recent book, *Freedom to Die*,

> A rational argument can be made for allowing PAS [physician-assisted-suicide] in order to offset the amount society and family spend on the ill, *as long as it is the voluntary wish* of the mentally competent terminally and incurably ill adult. There will likely come a time when PAS becomes a commonplace occurrence for individuals who *want* to die and feel it is the right thing to do by their loved ones. There is no contradicting the

> fact that since the largest medical expenses are incurred in the final days and weeks of life, the hastened demise of people with only a short time left would free resources for others. Hundreds of billions of dollars could benefit those patients who not only *can* be cured but who want to live. [Emphasis in the text.][37]

Should this crass attitude seep deeply into the culture, its stark utilitarian values could easily breach the levy of Hippocratic values and adversely impact normal health care decision making, further threatening the welfare of the elderly, disabled, and dying. Indeed, Humphry predicts that money will ultimately be the raison d'etre for legalized assisted suicide. The fiscal savings, he and his coauthor predict hyperbolically, will run into the "hundreds of billions of dollars."[38]

Logic is certainly on their side. With the advent of managed care, health care profits increasingly come from cutting costs. What could be more cost-effective than assisted suicide? On the micro level, imagine the potential impact on families whipsawed between providing end-of-life care and sending children to college. Most would do the right thing, but some would undoubtedly look to "death with dignity" as the stone that kills two birds.

This crass utilitarian equation has already been imposed on sick patients in Oregon under the state's rationed Medicaid law. In July 2008, Barbara Wagner was diagnosed with recurrent lung cancer. Her condition was considered terminal, but chemotherapy was an option to extend her life. When she asked Medicaid to cover the treatment, she received a letter from the administrator refusing to cover chemotherapy but offering to pay for her assisted suicide.[39] At about the same time, prostate cancer patient Randy Strop received a similarly heartless letter refusing chemotherapy but offering to pay for his assisted suicide.[40] Indeed, this is in keeping with the Oregon Health Plan's coverage of assisted suicide as "comfort care," but refusing to cover chemotherapy for patients who are very unlikely to survive for five years even if they receive treatment.[41]

Perhaps this is why poor people are not demonstrating in the streets demanding the right to assisted suicide: they are worried about receiving adequate care! Indeed, the terrible problem of poorer Americans obtaining fair access to medical treatment is one reason why LULAC, the largest Latino civil rights organization in the United States, has come out unequivocally in opposition to legalizing assisted suicide.

Beyond these concerns, we must also worry about the impact that the discrimination against the disabled and elderly, which already permeates society, would have on the actual practice of assisted suicide—regardless of the technicalities of the law that might (for a while) restrict access to assisted suicide to the terminally ill. In this regard, we should hearken to the clarion warnings and sage advice of the disability rights community.

Disability rights activists are the most committed and effective opponents of legalizing assisted suicide in the United States. This is highly significant given that disability rights activists tend to be liberal politically, pro-choice on abortion, and secular in their worldviews. Thus, like many assisted suicide opponents, their opposition to legalization isn't based on religion, but rather on their understanding that legalizing assisted suicide—even if formally restricted to the terminally ill—would validate discriminatory public attitudes that are profoundly threatening to people with disabilities.

Paul Longmore, a nationally respected disability rights activist, writer, and associate professor of history at California State University at San Francisco, has put it this way:

> Current euthanasia activists talk a lot about personal autonomy and choice. Well, for people with disabilities who have opted for assisted suicide, it was a spurious choice. These are people who have been denied the ability to choose about virtually every other option in their lives: They have been segregated out of society; they have been denied the right to work; they have been discriminated against in getting an education; they have been blocked from expressing themselves romantically and sexually; they have been penalized for marrying by having public benefits shut off, including desperately needed health insurance; they have been shunned by loved ones and friends. In virtually every case in which a person with a disability has sought legal assistance in ending their lives, they have been discriminated against in most if not all of these ways.[42]

Diane Coleman, one of the nation's most prominent disability rights activists and founder of Not Dead Yet, a disability rights organization that opposes legalizing assisted suicide and advocates for equal treatment for disabled people in the health care system, has expressed similar views:

> The widespread public image of severe disability as a fate worse than death is not exactly a surprise to the disability community. Disability rights activists have fought against these negative stereotypes of disability for decades in the effort to achieve basic civil rights protections. What has been a surprise for many advocates is the boldness with which these stereotypes are asserted as fact by proponents of assisted-suicide, and the willingness of the press and the public to accept them, without even checking them against the views of people who themselves live with severe disabilities. . . . These stereotypes then become grounds for carving out a deadly exception to longstanding laws and public policies about suicide prevention.[43]

These prejudices seep into the delivery of health care, undermining the quality of care received by the disabled. Examples of such biases are routinely reported in disability rights literature. The following excerpt from an article in the disability health and wellness journal *One Step Ahead's Second Opinion* is typical of the impediments placed in the paths of disabled people:

> Robert Powell has lived with partial paralysis since childhood and learned two years ago he has a heart condition. . . . [The] hospital staff repeatedly asked him how much he wanted done to save his life should his condition fail to respond to routine treatment. Having barely reached middle age, he assured them he wanted aggressive measures to save his life. Staff continued to question him about his decision. They finally requested a psychiatric consult because they felt he was "having trouble accepting death."[44]

Facing and overcoming this type of prejudice is difficult enough when the law values all lives equally. But with the assisted suicide movement now claiming that suicide should be sanctioned by the state in some cases, with the growing utilitarianism in bioethics, and in the context of society's prejudice against people with disabilities, many in the disability rights movement are convinced that they are the actual targets of the assisted suicide movement. Assisted suicide would be wrong even under the most ideal conditions. But in light of these stark realities, and considering the frayed safety net, legalizing assisted suicide in our country could have catastrophic consequences for the weak, vulnerable, depressed, and unwanted.

Drawing Conclusions

Based on the above, what can we conclude? First, the slippery slope is very real. The Dutch have proved that once killing is accepted as a solution for one problem, tomorrow it will be seen as the solution for hundreds of problems. Based on the ideology that undergirds assisted suicide advocacy, once we accept the assisted suicide of terminally ill patients, like the Dutch we will—over time—almost surely come to accept the facilitated deaths of other suffering people, perhaps even including the mentally ill. Indeed, we saw this slippery slope phenomenon in action during the nearly decade-long assisted suicide spree of Jack Kevorkian in the 1990s. Despite more than 70 percent of the people he assisted in suicide not being terminally ill—five weren't even ill upon autopsy—Kevorkian was (and in many ways remains) a popular figure.[45]

Second, adopting suicide as an acceptable answer to human suffering eventually changes popular outlooks. The law not only reflects our values, but in our diverse age, it tells us right from wrong. Accordingly, once suicide is redefined as medical treatment, it becomes transformed from "bad" into "good." Thus, the guidelines intended to "protect against abuse" eventually are viewed not as protections, but instead as hurdles separating sick and dying patients from the beneficence of death. (Indeed, the Kate Cheney case came to the *Oregonian*'s attention precisely because her family expressed "dismay that legal safeguards had become roadblocks to Kate's right to a lethal prescription."[46]) In such an intellectual and cultural milieu, it becomes all too easy to justify ignoring or violating "guidelines."

Third, legalizing assisted suicide can distort the attitude of medical professionals toward their sickest patients. Indeed, in the Netherlands, some Dutch doctors now refer patients who don't qualify legally for euthanasia to an approved how-to-commit suicide book, a process some Dutch doctors have called "autoeuthanasia."[47] This distortion of ethics and attitudes could become especially pronounced in a medical economic system dominated by cost containment and managed care where profits come from reducing the level of services.

Fourth, legalizing assisted suicide sends the implied message that people who are diagnosed with a terminal illness have lives less worthy of being protected than those of other suicidal people. Indeed, the Oregon law does not require suicide prevention when doctors are asked by a patient for suicide, while state officials have worried publicly about high suicide levels among the state's youth, elders, and other populations.[48]

****

In closing, let me quote the thinking of a wonderful friend, who I met when a hospice volunteer. Robert Salamanca of Pleasanton, California, died naturally of Lou Gehrig's disease. Bob believed strongly that assisted suicide advocacy diminished the value of the lives of people like him. Indeed, he felt so strongly about this that he wrote an important op/ed column in the *San Francisco Chronicle* (while completely paralyzed) that stated in part,

> Euthanasia advocates believe they are doing people like me a favor. They are not. The negative emotions toward the terminally ill and disabled generated by their advocacy is actually at the expense of the "dying" and their families and friends, who often feel disheartened and without self assurance because of a false picture of what it is like to die created by these enthusiasts who prey on the misinformed.
>
> What we, the terminally ill, need is exactly the opposite—to realize how important our lives are. And our loved ones, friends, and indeed society, need to help us feel that we are loved and appreciated unconditionally.[49]

Robert was right. The proper approach to death, dying, and other sources of human suffering is to increase our levels of care and compassion, not permit doctors to coolly write lethal prescriptions.

Moreover, whatever our beliefs might be about an idealized system of assisted suicide, in the real world of America today, legalizing assisted suicide would be dangerous and reckless. With our dysfunctional health care system; high rates of elder abuse; already alarming significant suicide levels; pronounced economic uncertainties; divisions of race, gender, religion, sexual orientation, class, and immigration status; and the concomitant lack of mutual trust, legalizing assisted suicide would be bad medicine and even worse public policy.

NOTES

1. International Task Force on Euthanasia and Assisted-Suicide, "Nine Years of Assisted-suicide in Oregon," based on 292 official reports on Oregon's "Death with Dignity Act," published annually by the Oregon Department of Human Services, http://www.internationaltaskforce.org/orrpt9.htm.
2. Ian Dowbiggin, *A Concise History of Euthanasia* (New York: Rowman & Littlefield Publishers, Inc., 2005), 50.
3. Rita Marker and Wesley J. Smith, "The Art of Verbal Engineering," *Duquesne Law Review* 35, no. 1 (1996): 81–107.
4. See, e.g., Kathryn L. Tucker and Fred B. Steele, "Patient Choice at the End of Life, Getting the Language Right," *The Journal of Legal Medicine* 28, no. 3 (2007): 305–325.
5. Derek Humphry, letter to the editor, *Register Guard*, November 8, 2006.
6. See, e.g., Derek Humphry and Mary Clement, *Freedom to Die: People Politics and the Right-to-Die Movement* (New York: St. Martin's Press, 1998), 30; and Elliot Cohen, Ph.D., "Permitted Suicide: Model Rules for Mental Health Counseling," *Journal of Mental Health Counseling* 23, no. 4 (October 2001), 279–294.
7. Jacob M. Appel, "A Suicide Right for the Mentally Ill? A Swiss Case Opens a New Debate," *Hastings Center Report* 37, no. 3 (May 2007): 21–23.
8. "Euthanasia Case Leeuwarden-1973," excerpts from court's decision, trans. Walter Lagerway, *Issues in Law and Medicine* 3 (1988): 429, 439–442.
9. Henk Jochemsen and John Keown, "Voluntary Euthanasia under Control? Further Empirical Evidence from the Netherlands," *Journal of Medical Ethics* 25 (1999): 16–21.
10. For more details on Dutch assisted suicide and euthanasia practices, see Wesley J. Smith, *Forced Exit: Euthanasia, Assisted Suicide, and the New Duty to Die* (New York: Encounter Books, 2006).

11. Kathleen Foley and Herbert Hendin, eds., *The Case against Assisted Suicide*, (Baltimore: Johns Hopkins University Press, 2002), 110.
12. Agnes van der Heide et al., "Medical End-of-Life Decisions Made for Neonates and Infants in the Netherlands," *The Lancet* 350 (1997): 251–255.
13. Astrid M. Vrakiing et al., "Medical End-of-Life Decisions Made for Neonates and Infants in the Netherlands, 1995–2001," *The Lancet* 365 (April 9, 2005): 1329–1331.
14. "No Prosecution for Dutch Baby Euthanasia," *Reuters*, January 22, 2005.
15. Eduard Verhagen and Peter J. J. Sauer, "The Groningen Protocol—Euthanasia in Severely Ill Newborns," *New England Journal of Medicine* 325, no.10 (March 10, 2005): 959–962.
16. J. Remmelink et al., *Medical Decisions about the End of Life*, Vol. I, Report of the Committee to Study the Medical Practice of Concerning Euthanasia and *Medical Decisions about the End of Life*, Vol. II, The Study for the Committee on the Medical Practice Concerning Euthanasia, the Hague, September 19, 1991 (hereafter cited as Remmelink Report I/II).
17. Remmelink Report I, p. 14, n. 2.
18. Richard Fenigsen, "The Report of the Dutch Government Committee on Euthanasia," *Issues in Law and Medicine* 7, no. 3 (Winter 1991): 340.
19. Remmelink Report I, 13.
20. Ibid., 15.
21. Remmelink Report II, p. 49, Table 6.4.
22. Ibid., p. 50, Table 6.6.
23. Ibid. 15.
24. This figure was arrived at by calculating 8.5 percent (the approximate percentage of all Dutch deaths that resulted from killing by physicians) of the total number of yearly U.S. deaths, which in 1985 was a little over 2 million (*New York Public Library Desk Reference*, 1989, 613).
25. Herbert Hendin, "Assisted Suicide, Euthanasia, and Suicide Prevention: The Implications of the Dutch Experience," *Suicide and Life Threatening Behavior* 25, no. 1 (Spring 1995): 202.
26. Ibid.
27. Witness testimony before House of Lords Select Committee on the Assisted Dying for the Terminally Ill Bill, published in House of Lords Select Committee, Vol. II: Evidence, April 4, 2005, p. 266, Q. 615.
28. Ibid., p. 363, Q. 592.
29. For example, see Timothy E. Quill, *Death and Dignity: Making Choices and Taking Charge* (New York: W.W. Norton & Co., 1994).
30. Kim Murphy, "Death Called 1st under Oregon's New Suicide Law," *The Los Angeles Times*, March 26, 1998.
31. Herbert Hendin et al., "Physician-Assisted Suicide: Reflections on Oregon's First Case," *Issues in Law & Medicine* 14, no. 3 (1998): 243–270.
32. Erin Hoover Barnett, "Is Mom Capable of Choosing to Die?" *The Oregonian*, October 17, 1999.
33. N. Gregory Hamilton and Catherine Hamilton, "Competing Paradigms of Response to Assisted Suicide Requests in Oregon," *American Journal of Psychiatry* 162 (June 2005): 1060–1065.
34. Ibid., 1062.
35. Ibid., 1063.
36. Ibid., 1064.
37. Humphry and Clement, *Freedom to Die*, 30.
38. Ibid.

39. Susan Harding, "Health Plan Covers Assisted Suicide but Not New Cancer Treatment," KATU, July 31, 2008, http://www.kval.com/news/26140519.html.
40. Dan Springer, "Oregon Offers Terminal Patients Doctor-Assisted Suicide Instead of Medical Care, Fox News, July 28, 2008, http://www.foxnews.com/story/0,2933,392962,00.html.
41. Oregon Health Services Commission, *2009 Prioritized List of Health Services*, page Sl-1.
42. Paul Longmore, as quoted in Smith, *Forced Exit* (2006), 193–94.
43. Diane Coleman, "Not Dead Yet," in Foley and Hendin, *The Case against Assisted Suicide*, 221.
44. Advanced Directives and Disability, *One Step Ahead's Second Opinion*, 2 No. 1, (Winter 1995), as cited in Wesley J. Smith, *Forced Exit: The Slippery Slope from Assisted Suicide to Legalized Murder* (New York: Times Books, 1997), 183.
45. L.A. Roscoe et al., "Dr. Jack Kevorkian and Cases of Euthanasia in Oakland County, Michigan, 1990–1998," *New England Journal of Medicine* 343 (December 7, 2000): 1735–1736.
46. Barnett, "Is Mom Capable of Choosing to Die?" *The Oregonian*.
47. Tony Sheldon, "Dutch Doctors Publish Guide to 'Careful Suicide,'" *British Medical Journal* 336 (2008): 1394–1395.
48. Associated Press, "Oregon Elder Suicide Rate Tops National Average, Study Says," published in the Corvallis *Gazette-Times*, August 24, 2005.
49. Robert Salamanca, "I Don't Want a Choice to Die," *San Francisco Chronicle*, February 19, 1997.

CHAPTER 13

CINEMATIC PERSPECTIVES ON EUTHANASIA AND ASSISTED SUICIDE

GLEN O. GABBARD

The cinema today has become the great storehouse for the psychological images of our time. It serves many of the same functions for contemporary audiences as tragedy served for the ancient Greeks in the fifth century B.C.[1] Not only do films provide catharsis, they also unite audiences with their culture and its mythological dimensions in the same way that Aeschylus or Sophocles provided meaning for the citizens of Athens centuries ago.

It can be argued that films do more than merely reflect the culture in which they were made. They also create that culture. In this regard, it is naïve to think that audiences make sharp distinctions between reality and what they see on the great silver screen. The cumulative effect of viewing film after film is the creation of a set of schema in our preconscious and unconscious memory banks. The laconic cowboy, the mad scientist, the whore with the heart of gold, and the socially inept intellectual are all examples of such internal stereotypes. Madison Avenue has long been aware of the power of such images. Manipulators of these images have used them to sell everything from toothpaste to political candidates. The cinema creates a set of myths in the culture. In the 1950s, John Wayne taught men how to be men, and Marilyn Monroe taught women how to be women. Indeed, when Tom Cruise appeared in the megahit *Top Gun* in 1986, enlistment in the navy's pilot program dramatically increased, even though what appeared in the film was pure fantasy.[2]

In the 1930s and 1940s, Nazi Germany appropriated this cinematic power in a series of films about eugenics, heredity, and euthanasia. These propaganda films were deliberately intended to promote an accepting attitude toward the necessity of killing in the Third Reich's master plan. In this chapter, I will illustrate parallels between the Nazi propaganda films and the way that euthanasia and physician-assisted suicide has been depicted in the commercial cinema of the last several decades. In brief, the parallels are disconcerting.

PROPAGANDA FILMS OF THE RWU

In the last 20 years, a number of films have been uncovered that were produced by the Reich Institute for Film and Images in Science and Education (RWU) and the Reich Office for Educational Film. These movies were ostensibly medical in nature and educational in purpose but clearly had a propagandistic agenda.[3] It is now clear that Hitler introduced a state-controlled educational film policy that was intended to politicize educational films and make the practices of sterilization, eugenics, and euthanasia acceptable to audiences who viewed the films so they could support the Third Reich programs in these areas.

Officials of the RWU had considerable powers to ensure that these films were seen. They were able to coerce professors in academic settings and schoolteachers to show the films to promote the political agenda of the Third Reich. The films, such as *The Curse of Heredity* (1927) and *Laws of Heredity* (1935), focused on those who were physically and mentally disabled. The films stigmatized these groups by claiming that these people had been allowed to survive and multiply in defiance of the laws of natural selection. Emphasis was placed on how expensive it was to maintain these unfortunate souls, and sterilization was promoted. Sweeping claims were made. One film proclaimed that if the physically and mentally disabled people were allowed to continue to procreate, one in four people would ultimately be "hereditarily ill." The need to kill the weak was presented with numerous examples, and many films concluded with stirring images of healthy German youth as counterpoint to the weak and the sick.

Existence Without Life, which was finished in 1941, features a professor identifying mental illness as a hereditary evil. Distorted faces made to look grotesque by the lighting are shown on the screen while the professor describes the individuals as incurable. The professor, whose apparent "authority" is intended to provide a touch of gravitas, uses the concept of "mercy" as a way of dealing with them to relieve them of their existence without life. He ends the film with a powerful personal comment: if he were afflicted with such a condition, he would do anything to prevent it. He even goes so far as to say that he would rather die and that he has no doubt that the mentally ill feel the same. The professor then suggests that there is an obligation to allow these people to exercise their rights. He regards them as victims of a horrible fate, who can neither get better nor die. The 1941 film *I Accuse* goes even further in deliberately conflating the ethical issue of voluntary euthanasia with state-sanctioned mass murder.[4] This film, seen by over 15 million people, was directed by one of the outstanding filmmakers in Germany, Wolfgang Liebeneiner, and features a pianist struck with multiple sclerosis as a central character. She appeals to her doctor to kill her, arguing that she is no longer a person, but rather a lump of flesh. When her doctor refuses, her husband takes her life and is then shown in court, where he is asked about the murder. He replies that his reason for killing her is that he loved her. The narrative of the film clearly tries to convince audiences that courts need to allow more flexible laws regarding euthanasia.

During the 1930s, the whole educational structure was modified according to Nazi ideology. A leading bureaucrat, Kurt Zierwold, felt strongly that the power of film as a mass medium should be exploited to educate the public about the aims of national socialism. He suggested that propaganda geared to a small number of people did not make sense and that film had the capacity to reach the masses.[5]

Throughout all these films, a recurrent theme is that those with serious mental illness or incurable physical disease are violating the laws of nature and would certainly die if left to natural forces. In the 1935 film *The Laws of Heredity*, sick and weak animals are seen dying, and these shots are juxtaposed with depictions of psychiatric patients in an asylum.

We can only speculate about the impact such films had on the citizens of Nazi Germany. We know, however, that hundreds of physicians willingly participated in the "euthanasia" inherent in genocide.

EUTHANASIA THEMES IN HOLLYWOOD AND EUROPEAN CINEMA

We will now consider the depiction of euthanasia in more contemporary films. A broad definition of euthanasia is applied as we consider these films. All instances in which deliberate actions are taken with the intent of ending a patient's life, whether they are assisted or are actually carried out by a family member or health professional, fall under this category. Within this definition, it does not matter if a doctor or the patient administers the drug since the legal differences are subtle.

If we examine the popular films of the last four decades, we see a clear message—euthanasia is a compassionate, rational choice made by the patient alone, and it is unreasonably opposed by insensitive, misinformed, and unenlightened mainstream institutions, including the Catholic Church, the courts, medicine, and psychiatry. In many of these movies, the protagonist must fight to die on his or her own terms. Table 13.1 depicts a selective list of motion pictures with a euthanasia theme, however minor.

One would be surprised to think that films involving euthanasia are box office hits because the topic would seem to be unpopular with audiences, given the general aversion to death that is rampant in contemporary Western culture. Nevertheless, it is striking how many of the films in table 13.1 have been honored with Best Picture Oscars. *The English Patient*, *Million Dollar Baby*, and *One Flew over the Cuckoo's Nest* all won Academy Awards for Best Picture, and *The Sea Inside* and *The Barbarian Invasions* won the Oscar for Best Foreign Film.

Table 13.1 Films from 1970 to 2006 Showing Variations of Euthanasia

Johnny Got His Gun (1971)	*The English Patient* (1996)
Harold and Maude (1971)	*It's My Party* (1996)
Soylent Green (1973)	*One True Thing* (1998)
Murder or Mercy (1974)	*Igby Goes Down* (2002)
One Flew over the Cuckoo's Nest (1975)	*Talk to Her* (2002)
An Act of Love (1980)	*The Event* (2003)
Whose Life Is It Anyway? (1981)	*The Barbarian Invasions* (2003)
Right of Way (1983)	*The Sea Inside* (2004)
Grace Quigley (1985)	*Million Dollar Baby* (2004)
Last Wish (1992)	*Just Like Heaven* (2005)
	Children of Men (2006)

Why audiences and Academy voters celebrate these films is not entirely clear. An examination of the themes that emerge from these films provides us with some clues. Successful films often tap into commonly held unconscious wishes and fears in the mass audience.[6] We go to the darkened theatre to master shared and universal anxieties. The cinema is mythopoetic—it creates myths that we need to get through the existential dread associated with living. As Claude Levi-Strauss stressed, myths are transformations of fundamental conflicts or contradictions that in reality cannot be resolved.[7] This applies in particular to our most fundamental anxiety: death. Mercy killing becomes a simple, obvious, uncomplicated solution to suffering, pain, family anguish, and the fear of the great unknown.

A Simple Choice

As suggested above, many of these films depict the wish to die as a rational choice made by a sound mind when life is no longer worth living. In John Badham's 1981 film, *Whose Life Is It Anyway?*, Richard Dreyfuss plays a sculptor who has a serious auto accident, rendering him quadriplegic. As he lies in a hospital bed with no prospect of a life in which he is able to sculpt or to make love, he chooses to die. He must fight a conservative hospital staff and a traditional court system to convince them that he should be allowed to die, but the film clearly argues that he is making a reasonable decision. In his final address to the patient, the judge (Ken McMillan) says he is satisfied that the quadriplegic patient is not suffering from depression and is able to make a decision that must be respected.

This film, like others, serves to generate a particular view about terminally ill or chronically disabled patients—namely, that most of them wish to die as we would if *we* were in their situation. Research suggests otherwise. Harvey Max Chochinov et al. studied the desire for death in a group of 100 cancer patients who were terminally ill.[8] Only 8.5 percent harbored a clinically significant desire for death, and of those, 59 percent met the criteria for major depression. Similarly, in a 1996 study of 300 men with AIDS, there was no association between severity of illness and the wish to die.[9] Steven H. Miles points out that doctors who treat terminally ill patients may have a "pseudoempathy" in which they view the patient's mood state as basically normal, based on the assumption that that's the way they would feel if they were in a similar situation.[10] As a result, clinical depression may go undiagnosed and untreated.

The quadriplegic patient in *Whose Life Is It Anyway?* is shown as making a decision to deny any form of heroic treatment—not out of depression, but out of a sense that he has an existence without life. The judge takes a psychiatrist's word that he is not depressed and, apparently, does not consider that the accident is fairly recent and that the patient may be in a state of grief that will ultimately lift with the passage of time and adjustment to the illness. In one study, half of terminally ill cancer patients had changed their minds over a two-month period.[11]

Two of the most romanticized depictions of mercy killing are in films like *The Barbarian Invasions* (2003) and *The Sea Inside* (2004), where the act of assisting a terminal family member commit suicide is seen as the ultimate act of love. In Alejandro Amenábar's *The Sea Inside*, Ramon (Javier Bardem) has become quadriplegic as a result of a diving accident. He sees no reason to continue living. He lies in bed and has fantasies of flying. He finally tells his girlfriend, "The one who loves you is the one who kills you." They are shown lovingly kissing at the end of the film and watching the sun set prior to his death.

In Denys Arcand's *The Barbarian Invasions*, Remy (Remy Girard) is a divorced man in his 50s hospitalized with cancer. His son, Sebastian (Stephane Rousseau), from whom he has been estranged, comes from London to Montreal to help him through his death. Sebastian goes to great lengths to ease his father into death. The tearjerking scene where they say goodbye to each other once again reflects that his son's willingness to participate in euthanasia is the ultimate act of love.

The depiction of the decision to kill one's family member in both of these films has a chilling similarity to some of the Nazi propaganda films made by the Third Reich. If one watches the 1941 film *I Accuse* and listens to the dialogue between the husband and his wife with multiple sclerosis, the words are eerily similar. In *I Accuse*, the filmmaker deliberately blurs the distinction between voluntary and involuntary euthanasia. An assumption is made that anyone in such a debilitated state would want to die, and the empathic, compassionate thing to do is to give them their wish. Similarly, films like *The Sea Inside* and *The Barbarian Invasions* fail to depict any hint of coercion by the family or others, suggesting that the patients themselves were unambivalent in their pursuit of being put to death.

When the Netherlands allowed euthanasia to be legal under certain restricted circumstances, Dutch psychiatrists noted that subtle or not-so-subtle coercion went on between the family of the patient and the terminal patient. At times, there appeared to be outright pressure brought to bear by family members who would say to the aging parent that perhaps their life was no longer worth living. Coercion can influence a decision about euthanasia or physician-assisted suicide at any point in the process. It can affect when to request the overdose, whether or not the overdose is taken, and how long the patient chooses to live between making the request and overdosing.[12]

Nathan Cherny et al. coined the term *empathic distress* to describe how dying patients may perceive the distress of their family and health care providers.[13] This perception further compounds their own distress and may make them feel that their existence constitutes a burden to others. This sense of family fatigue may lead patients to request assisted death out of consideration for their loved ones. These complex family dynamics are missing from films that depict loving instances of euthanasia.

Similarly, the view that the patient himself decides on assisted suicide may overlook how influential physicians are in the decision. Edmund Pellegrino has argued against the institution of assisted suicide because it often falsely implies that the initiative is in the patient's hands, when in fact the provision of means and the clinical evaluation are often in the hands of the doctor.[14] Herbert Hendin cautions that physicians may help the patient make the decision because they, too, are overwhelmed with their powerlessness and helplessness in the face of a steadily deteriorating patient.[15] Physicians prefer to harbor the illusion of mastery over disease, and the patient takes that away from them. The rapid assumption that a request for physician-assisted suicide is what it appears to be can be highly problematic. Zealots for the procedure may automatically take what the patient says at face value without further exploration. Studies indicate that the request to die can be a plea for help or a hope that the doctor can provide a reason to live. Moreover, it can be a way of indirectly asking for more aggressive palliative care. Still other patients are asking for reassurance about their value.[16] In addition, the doctor's countertransference anxiety or helplessness may lead to a rapid decision without proper exploration of the underlying meaning of the request for euthanasia. In one survey, 44 percent of the physicians admitted that they experienced the anxiety of a terminally ill patient as unbearable sometimes.[17]

In Carl Franklin's *One True Thing* (1998), Meryl Streep plays Kate Gulden, a saintly mother who bravely faces her demise from cancer. Her philandering husband, George (William Hurt), distances himself from her as she dies, while her daughter, Ellen (Renee Zellweger), returns home to take care of her. The film is initially ambiguous regarding the exact explanation of Kate's death. At one point, the mother desperately asks her daughter to help her take her life because her life is no longer worth living. Here the film resonates with the Nazi propaganda films. However, after her death, neither the audience nor the characters in the film are clear on how she died. In a moving cemetery scene, the daughter and the father confess that they both assumed the other one had given Kate a fatal overdose. They both express dismay as they assert that they had nothing to do with her death. They suddenly realize that she took the overdose on her own volition. With profound admiration and respect, her husband says, "Of course she did it. Who else would have had the strength?" The act of ending her life is clearly a heroic act that only an incredibly strong person could carry out. Indeed, the film suggests that the husband and the daughter were neglectful in not doing it themselves. There is no virtue, the film suggests, in tolerating the pain and letting nature take its course, as the dying family member is supported and treated with palliative analgesics. This is the coward's way out.

THE OUTLAW HERO

Another theme that emerges from a careful study of the films that endorse euthanasia is the myth of the outlaw hero, pervasive in the American cinema. One such hero is Frankie Dunn (Clint Eastwood) in *Million Dollar Baby* (2004). In this film, directed by Eastwood himself, Frankie is a gruff boxing trainer who reluctantly coaches Maggie Fitzgerald (Hillary Swank) to boxing fame. After assisting her in her journey to boxing stardom, he watches Maggie's arch nemesis break her neck with an illegal maneuver in the boxing ring.

After her spine injury, Maggie is confined to bed with little ability for movement, and she clearly no longer wishes to live. Frankie is wracked with agony about assisting her in her wish to die. Frankie visits his hard-boiled priest (Brian O'Byrne), with whom he has a tenuous relationship. The priest admonishes him in no uncertain terms that no matter how many sins he has committed earlier in his life, it would be an unforgivable sin in the eyes of God to end Maggie's life in an apparent act of compassion. Frankie shuns the advice and compassionately puts his young mentee to sleep with a lethal injection after disconnecting her airway. His beautifully acted, heartfelt farewell leaves the audience in tears.

The casting of Clint Eastwood in the role of Frankie is significant in this film because his iconic image on the screen resonates with a long history of outlaw heroes, notably Dirty Harry Callahan. In the Dirty Harry films, it is clear to the audience that the legal system has failed to bring justice to a vile killer. Harry Callahan reluctantly becomes the outlaw hero who takes the law into his own hands because he has no other choice if he wishes to see justice. Just as the court system is impotent in those films, in *Million Dollar Baby*, the Catholic Church is the institution that has become hopelessly insensitive and entrenched in dogma. The priest is seen as a rigid adherent to an outmoded theology that lacks compassion. In *Dirty Harry*, Callahan thumbs his nose at the law. In *Million Dollar Baby*, Frankie Dunn thumbs his nose at the Church. The audience derives great satisfaction from these scenes in which the outlaw hero acts according to his own conscience, which is guided by a higher principle than those of the mainstream establishment institutions.

Audiences have always loved the outlaw hero in traditional American cinema. Popular American films frequently avoid answering the most troubling questions by displacing them onto melodrama, where solutions are more easily found. Many of these films shift the center of gravity involving major social themes to the decisions of a single individual. Hence, American movies reinforce the often paradoxical belief that the answer, and ultimately the solution, to just about anything lies within ourselves, not within the society's institutions.

EUTHANASIA AS PUNISHMENT FOR ONE'S SINS

In Anthony Minghella's *The English Patient* (1996), Count Laszlo de Almasy (Ralph Fiennes) is badly burned in a plane crash. He is a Hungarian mapmaker charting the Sahara Desert in the 1930s. However, with the onset of World War II, he falls in love with a married woman (Kristin Scott Thomas) whom he cannot resist. The star-crossed lovers are stranded in the desert, and Almasy leaves his beloved in a cave while he crosses the desert seeking help. He cannot move her because of severe fractures she sustained in the plane crash. She dies in the cave because the count is unable to rescue her.

As he lies in bed, consigned to unrelenting pain under the care of the compassionate nurse Hana (Juliette Binoche), the count is wracked with guilt over what has transpired. He says he cannot be killed because he has been dead for years. Near the end of the film, he knocks over a box of opiate vials and pushes them, one at a time, toward Hana as she prepares his injection. Through her tears, she knowingly overdoses him by injecting each vial into his intravenous line. He says, "Thank you," to her in a barely audible whisper. She reads to him at his request, and he imagines his beloved as he breathes his last breath. The film then cuts to a shot of a plane ascending from the desert, symbolically reflecting his flight into heaven.

In this slant on the euthanasia theme, at one level the count is being relieved of his pain. In his own mind, however, he is also being executed as punishment for his neglect of his true love.

THE OTHER SIDE OF THE STORY

It is ironic that one must search far and wide among the films that deal with euthanasia to find one that raises questions about the practice and that treats the issue with greater complexity. It is perhaps even more startling to note that a rather ordinary romantic comedy, Mark Waters's *Just Like Heaven* (2005), is the film that meets these criteria. Reese Witherspoon plays Elizabeth Masterson, a workaholic doctor who is injured in a car crash and lapses into a deep coma. A struggling artist, David Abbott (Mark Ruffalo), moves into her San Francisco apartment. However, he is shocked to find that Elizabeth's spirit, which only he can see, haunts the apartment. They gradually fall in love, only to learn that Elizabeth's relatives have decided to disconnect life support. There is an intriguing scene in which the doctor (Ben Shenkman) emphasizes that he is acting on an advance directive that Elizabeth has signed, in which she states that she does not wish to have her life continued on life support. Elizabeth, speaking as a ghost (which means the doctor cannot hear her, but David can), notes that she has since changed her mind, and she does not want the life support to be disconnected. Through heroic measures, David manages to avoid this passive euthanasia, and Elizabeth arises from the coma, much like Sleeping Beauty, who is brought to life by Prince Charming. The film clearly suggests that it would have been a profound tragedy if the doctor had succeeded with his plan to disconnect the life support.

Indeed, research suggests that the movie has a point. The study of 100 terminally ill cancer patients (noted above) by Chochinov et al. found that the desire for death is temporally unstable.[18] Some patients who want to die at one point change their minds with the passage of time, raising considerable concern about responding to a patient's requested wish to die at any particular point in the disease. In fact, in the Netherlands, when adequate pain alleviation was provided, 85 percent of patients withdrew their requests for euthanasia or physician-assisted suicide.[19] Similar findings have been noted in Oregon.

CONCLUSION

As is often the case, the cinema provides an alternative reality for audience members and provides a mythic solution to an extraordinarily complicated dilemma. Many of the films covered in this chapter are beautifully made and emotionally moving. The patient who is helped to die generally passes into a painless sleep as his loved ones look on. In the movies, one rarely sees complications, such as vomiting, that have occurred in Oregon cases, or the lag of several hours between administration of the lethal agent and the death. In one well-documented case, a patient took the overdose, lost consciousness in 25 minutes, and then regained consciousness 65 hours later.[20]

Of even greater concern is the "slippery slope" phenomena that occur when a society begins to endorse euthanasia as a sanctioned means of dealing with terminal illness. As horrifically depicted in the Nazi films of the 1930s and 1940s, it is a relatively small step to "selections," a God-like determination of who is worth saving, and a misperception of where choice resides. Even in the Netherlands, where they have been careful to institute a series of safeguards prior to reaching a decision on euthanasia, the debate has now turned to children who are born with serious defects and are unlikely to live more than a few years of childhood. Should they be allowed to die shortly after birth?

Finally, the cinematic portrayals of euthanasia and assisted suicide may provide us with a misleading sense of mastery over the existential dread associated with death. One way to view these films is that they displace the horror of death onto the means of dying, which provides some reassurance about the audience's anxiety in the face of death. These films suggest that one can arrange the moment of one's death with the loving help of family members, and in so doing, death is less likely to surprise us or terrorize us.

NOTES

1. Glen O. Gabbard and Krin Gabbard, *Psychiatry and the Cinema, 2nd Edition* (Washington, D.C.: American Psychiatric Press, 1999).
2. Gabbard and Gabbard, *Psychiatry and the Cinema.*
3. Ulf Schmidt, *Medical Films, Ethics, and Euthanasia in Nazi Germany* (Husum, Germany: Matthiesen Verlag, 2002).
4. Ibid.
5. Ibid.
6. Glen O. Gabbard, "The Psychoanalyst at the Movies," *International Journal of Psychoanalysis* 78 (1997): 429–434.
7. Claude Lévi-Strauss, *The Raw and the Cooked: Introduction to a Science of Mythology: I*, trans. John and Doreen Weightman (New York: Harper and Row, 1975).
8. Harvey Max Chochinov et al., "Desire for Death in the Terminally Ill," *American Journal of Psychiatry* 152 (1995): 1185–1191.
9. William Breitbart, B.D. Rosenfeld, and Steven D. Passik, "Interest in Physician-Assisted Suicide Among Ambulatory HIV-Infected Patients," *American Journal of Psychiatry* 153 (1996): 238–242.
10. Steven H. Miles, "Physicians and their Patient Suicides," *Journal of the American Medical Association* 271 (1994): 1786–1788.
11. Chochinov, "Desire for Death in the Terminally Ill."
12. Madelyn Hsiao-Rei Hicks, "Physician-Assisted Suicide: A Review of the Literature Concerning Practical and Clinical Implications for UK Doctors," *BMC Family Practice* 7 (2006): 9–17.
13. Nathan I. Cherny, N. Coyle, and K.M. Foley, "The Treatment of Suffering when Patients Request Elective Death," *Journal of Palliative Care* 10 (1994): 71–79.
14. Edmund D. Pellegrino, "Compassion Needs Reason, too," *Journal of the American Medical Association* 270 (1993): 874–875.
15. Herbert Hendin, "Seduced by Death: Doctors, Patients, and the Dutch Cure," *Issues in Law and Medicine* 10 (1994): 123–168.
16. Hicks, "Physician-Assisted Suicide."
17. Maria A. Annunziata et al., "Physicians and Death: Comments and Behavior of 605 Physicians in the Northeast of Italy," *Supportive Care in Cancer* 4 (1996): 334–340.
18. Chochinov, "Desire for Death in the Terminally Ill."
19. P.J. Van der Maaset et al., "Euthanasia and other Medical Decisions Concerning the End of Life," *Lancet* 338 (1991): 669–674.
20. Hicks, "Physician-Assisted Suicide."

CHAPTER 14

SCIENCE, MEDICINE, AND RELIGION IN AND AFTER THE HOLOCAUST

JOHN M. HAAS

Science, medicine, and religion are powerful social forces. The Hippocratic Oath itself, the foundational document of ethical medical practice, calls upon deities to witness to the taking of it: "I swear by Apollo the Physician and Asclepius and Hygieia and Panaceia and all the gods and goddesses making them my witnesses, that I will fulfill according to my ability and judgment this oath and this covenant."

In the Catholic Church, one can see this natural and enduring relationship between religion and medical science. The Church has a very long history with health care. From the earliest days, following the example and the teachings of Jesus, Christians would take the sick and the infirm into their homes, which eventually became hospitals. From the time of the establishment of a Christian social order in the fourth century, hospitals, asylums, leper colonies, and orphanages were run by religious communities in the Church.

This Catholic tradition of the melding of religion and health care continued in this country when large numbers of Catholics immigrated to the United States. Catholic immigrants were often denied health care in other non-Catholic institutions, or they were so poor they simply could not afford it. So communities of religious sisters, brothers, and priests started hospitals, orphanages, and asylums. Such initiatives have continued to grow so that in our day the Catholic Church is the largest nongovernmental provider of health care in the United States, generating $90 billion a year in revenue and holding over $200 billion in assets.

Furthermore, Catholics involved in care for the sick and dying had no problem embracing the pagan Hippocratic Oath, committing themselves never to violate a patient entrusted to their care. One can find early manuscripts with this pagan oath written in the form of a cross. And our last pope, John Paul II, in one of his formal teaching documents, called an encyclical, referred to "the still relevant Hippocratic

Oath, which requires every doctor to commit himself or herself to absolute respect for human life and its sacredness."[1]

It is this uncompromising commitment to the dignity of each living human being that can save health care from ever being put to sinister and disordered purposes, as has happened more than once in history. In the Germany of National Socialism, everyone and everything was ordered for the good of the state, not of the individual. Indeed, if need be, the individual was willingly sacrificed for the sake of the state. There ceased to be protections of the dignity of the human person.

To gain a better understanding of how the medical profession could have become so corrupted by National Socialism, one must place it in a larger historical context. The late nineteenth century had seen the rise of eugenics movements not only in Germany, but in the United States and England as well. A corruption of the medical profession had actually begun well before Hitler's rise to power. A movement known as Social Darwinism had spread throughout Europe and America. It was thought that nations as well as species were engaged in a struggle that led to the survival of the fittest. The fittest nations would be those that were genetically most pure and robust.

Indeed, as early as 1865, the eugenics movement had been initiated by Sir Francis Galton, a cousin of Charles Darwin. It drew many prominent individuals such as George Bernard Shaw, Alexander Graham Bell, and Margaret Sanger, the founder of Planned Parenthood. International Eugenics Conferences were held in London in 1912 and in New York in 1921 and 1932.

After the First World War, an association developed between eugenicists in the United States and England and so-called racial scientists in Germany. Fritz Lenz developed his theories of racial inequality and complained that the constitution of the Weimar Republic restricted the means that could be used to reduce the numbers of "undesirables." He pointed to the advanced eugenics programs in the United States and England and argued that involuntary sterilizations would be one way to avoid the growth of the racially degenerate elements of the population.[2]

It is clear that the Jews were not the only ones subjected to National Socialists racist theories that were a natural, though abhorrent, outgrowth of Social Darwinism. The degenerate races included the Africans, Gypsies, and Slavs. These came to be known as the *Untermenschen*, the "subhumans." But even here, we find that the concept was not isolated and hardly exclusive to the National Socialists in Germany.

The first use of the term *Untermenschen* was by an American, Lothrop Stoddard, in a 1922 book that consisted of a tirade against the degenerate Slavs who had taken power in the Soviet Union as the Bolshevists. It was entitled *The Revolt against Civilization: The Menace of the Under Man* and was published by the reputable publishing house of Charles Scribner's Sons in New York. It argued that if the white race was going to win the struggle against the inferior races, it must eschew liberal policies and adopt an expansive eugenics program. His book was published in Germany in 1925 as *Der Kulturumsturz: Die Drohung des Untermenschen,* and the term and concept of the *Untermensch* were soon taken up by Nazi apologists.

As we reflect on the corruption of the medical profession by National Socialism, it is dismaying to note that the professional group that comprised the highest number among the National Socialists were, indeed, the physicians. That may be attributable in part to the fact that National Socialism had a sort of "pseudoscience" as an integral part of its political and international program that grew out of the eugenics movement. As Dr. E. Ernst points out, "Hitler frequently referred to the Jewish race in medical terms, as 'bacillus,' 'parasite,' or 'disease.' His followers adopted these medical analogies. In their minds, the 'biological body of the German people' *(Volkskörper)* was threatened by the Jews."[3]

As has been pointed out, even before Hitler came to power, the notions of racial purity and racial health were widely accepted in the medical and scientific communities. These ideas could, therefore, resonate with the crowds as when Hitler spoke at a party rally as early as 1929 in Nuremberg. At the rally, he railed that the people must work for the "future maintenance of our ethnic strength *(Volkskraft)*, indeed of our ethnic nationhood *(unseres Volkstums)* altogether."[4]

In fact, a German document in 1935 declared, "The maintenance of a genetically homogeneous and healthy stock of the German people is the main duty of the German physician."[5] Race hygiene had been developed by, and had come under the purview of, the medical profession. As Ernst puts it, "To a large degree, the medical profession was not politicized but politics were medicalized."[6]

It is noteworthy that there were inevitably attempts to provide the appearance of legality and due process of law for even the most inhumane and unjust actions, such as the 1933 Sterilization Law and the 1935 Nuremberg Laws. As abhorrent as these laws were, as Catholics we believe that something good can be seen here. We hold to a theory, if you will, known as the "natural moral law." We believe that there is a moral law that is written by God in the hearts of all human beings so that when that moral law is violated, there is always an awareness of the violation at some level. Why else would the Nazis have devised the appearances of due process and legality to justify unspeakably inhumane behaviors unless at some level they recognized their immorality? There was a Jewish Pharisee two millennia ago, Saul, also called Paul of Tarsus, who wrote a letter to the Romans and mentioned this reality: "For when the Gentiles who do not have the law by nature observe the prescriptions of the law they show that the demands of the law are written on their hearts."[7] In fact, we would maintain that the attempt on the part of the National Socialist and other criminal political and social movements to justify their behaviors in terms of morality is an attestation to this law. Indeed, it is the basis of international law and provided legitimacy for the Nuremberg Trials after the war and judgments of crimes against humanity.

In October 1939, Hitler quietly established a broad euthanasia program to dispose of children with Down syndrome, hydrocephaly, and other anomalies. He wrote, backdating his order to September 1, the day the war began: "Reichsleiter Bouhler and Dr. med. Brandt are commissioned with the responsibility of extending the authority of specified doctors so that, after critical assessment of their condition, those adjudged incurably ill can be granted mercy-death."[8]

By the end of the year, the euthanasia program was extended to adults suffering from "lives not worth living" *(das lebensunwerte Leben)*. However, despite Hitler's written order, the program was still being carried out surreptitiously, and most of the general population knew nothing of it. Physicians had generally repudiated euthanasia during the Weimar Republic. However, it was psychiatrists who became increasingly open to eliminating the mentally ill. The general public, on the other hand, would not have accepted it. Even as late as 1939, the Nazis were aware that the public was not prepared to accept euthanasia, particularly those who were practicing Christians. Hitler quite consciously avoided introducing a program of euthanasia throughout the 1930s because he did not want to provoke the condemnation of the Catholic Church.[9] Here we can see the value of having competing or complementary institutions within a society, providing some sort of checks and balances. Here we can also see the value of a certain pluralism in the social order. In this case, there was a brutal dictatorship that did not dare undertake certain courses of action because of its fear of losing its hold on power or because of its fear that it could not effectively carry out its programs.

Nonetheless, despite general public disapproval, there had been a slow and incremental acceptance of euthanasia among the social elite. As early as 1933, the Prussian minister of justice, Hanns Kerrl, declared that *voluntary* euthanasia, certified by two physicians, would not be a criminal offense. This is similar to what is permitted in the United States today in Oregon and Washington, which have legalized physician-assisted suicide. Kerrl also stated that it would not be a criminal offense to eliminate the incurably mentally sick; that is, it would not be a criminal offense to inflict *involuntary* euthanasia on those who were not dying! As the Hitler biographer Ian Kershaw points out, "The hierarchy of the Catholic Church responded in predictably hostile fashion."[10]

By the beginning of 1941, rumors had become increasingly widespread that the Nationalist Socialist regime was engaged in a euthanasia program in the asylums throughout the country. On August 3 of that year, Bishop Clemens August Count von Galen issued a pastoral letter and denounced the practice of euthanasia in blunt and courageous terms: "There is a general suspicion, verging on certainty," the bishop insisted, "that these numerous unexpected deaths of the mentally ill do not occur naturally but are intentionally brought about in accordance with the doctrine that it is legitimate to destroy a so-called worthless life—in other words, to kill innocent men and women if it is thought that their lives are of no further value to the people and the state."[11] In this sermon, von Galen spoke right to the undermining of the medical profession and the violation of the physician-patient bond that was occurring. "Who could then have any confidence in a doctor? He might report a patient as unproductive and then be given instructions to kill him! It does not bear thinking of, the moral depravity, the universal general mistrust which will spread even in the bosom of the family if this dreadful doctrine is tolerated, accepted, and put into practice." Von Galen went on to say, "We will keep ourselves and our families out of reach from their [Nazi] influence, lest we become infected with their ungodless ways of thinking and acting, lest we become partakers in their guilt and thus liable to the punishment which a just God must and will inflict on all those who . . . do not will what God wills."

Following this homily, von Galen suffered virtual house arrest until the end of the war. His life was spared, however, because Hitler feared an uprising of the Germans in Westphalia if the bishop were hanged as Bormann had urged. In his sermon, von Galen had carefully woven three themes together, all of which had unsettled Catholics and threatened to undermine their support, which Hitler needed at this point in the war. The bishop decried the emptying of the monasteries and appropriation of their property, which had also occurred in 1941, and he judged the bombing of German cities to be a punishment of God for the killing of the innocent in the rumored euthanasia programs taking place in the asylums.

The homily had its effect. Hitler himself was given a copy of it. Thousands of secret copies of the sermon were distributed hand to hand. As Kershaw puts it, "The secrecy surrounding the 'euthanasia action' had been broken," and the secrecy was broken by an institutional source of authority that could, to a certain extent, act as a counterweight to the state.[12]

Worried that popular support for the war effort had been seriously weakened and wanting to avoid a confrontation with the Church, Hitler stopped the Euthanasia Program on August 24 as quietly and as secretively as he had started it. Even so, the National Socialists used the opportunity to declare Christianity and Nazism to be incompatible. They wrote,

> It is clear that the Christian churches try to keep alive even creatures unworthy of life, those who are totally and incurably mentally ill. But it is equally clear that we National Socialists must see things from a different point of view.... To keep them alive is completely against nature, because by nature they are in no way enabled to stay alive on their own.[13]

Regrettably, however, the killing continued, but no longer as part of the sanctioned Euthanasia Program. Between 1941 and 1945, tens of thousands of individuals in concentration camps were selected, again regrettably by physicians, as being too ill or weak to work and were subsequently put to death. Those physicians who had gained particular expertise in gassing techniques were deployed to the East. Now that the Euthanasia Program had been brought to an end, their expertise was put to use in a much larger program, the extermination of the Jews and other so-called *Untermenschen* of Europe.

Germany, and particularly the German medical profession, continues to live under the shame and ignominy of the National Socialist period. If we would speak of medicine, science, and religion after the Holocaust, one must acknowledge that Germany has some of the most rigorous ethical guidelines for medicine and research in Europe today. It must also be acknowledged that this is due in no small part because of its sad, recent history. The medical establishment has sometimes been criticized for not being willing to reflect more deeply in the postwar period on the role of their own profession in the crimes of the National Socialist period. However, physicians, politicians, and the Church in Germany are today committed to avoiding the kinds of horrors that occurred under the National Socialists. Research on human subjects is very carefully monitored.

Yet it must be pointed out that some of the legal sanctions Germany has in place against what it considers to be practices in violation of human dignity are considered too rigorous by other member states of the European Union. Germany, for example, has a law protecting the embryo and has outlawed Preimplantation Genetic Diagnosis (PID). In this procedure, an embryo that has been engendered through in vitro fertilization will have a single cell removed that is then tested for a genetic defect. If the embryo is carrying any genetic defect, no attempt is made to have it implanted, and it is destroyed. However, memory of the eugenics program of the National Socialists was strong and painful enough that there has been strong opposition to this procedure in Germany.

In a 2001 speech in Berlin, Johannes Rau, the president of Germany, addressed the controversial procedure. "Many demand," said the president, "that Preimplantation Genetic Diagnosis, PID for short, should be allowed also in Germany. The issue here is whether an embryo engendered through IVF should be identified by genetic defect before it is placed in the mother's womb. May the embryo be set aside or should it be devalued when such defects are determined?" He added that Germany should protect itself against the practice that most likely would lead to biological selection, that is, to a type of eugenics.

He went on,

> Children are a gift. I know how bitter it is for many who have no children. When there is the possibility to engender children artificially or to test the genetic composition of an embryo, then will there not arise the opinion that . . . one has the right to healthy children? Wherever there are unfulfilled wishes, and it appears that they can be fulfilled, then there quickly arises the appearance of a right.
>
> We know, however, that there is no such right. As understandable as such desires and wishes are, they do not constitute rights. There is no right to children. There is, however, a right of children to loving parents, and above all the right to come into the world and to be loved for their own sakes.[14]

In speaking of the raging bioethical debates of our day he said,

> No one should be surprised that the churches are particularly engaged. It would be an error to believe therefore that it has to do merely with a sectarian morality unique to the churches. One hardly has to be a believing Christian to know and to sense that certain possibilities and proposals for the biological and genetic technologies contradict fundamental values of human life. This value system has been developed by a history stretching back millennia—and belongs not just to Europe. They rest on the simple assertion, enshrined in our constitution, that the worth of the individual person is inviolable.[15]

In the remarks of Johannes Rau, one can see the confluence of medical, scientific, and religious concerns in the post-Holocaust period in Germany. In his remarks, he appealed obliquely to Hippocrates, to the natural moral law, and to the religious institutions in German society. He appealed openly to the concerns of the members of society. There is the hope that all of these factors, all of these institutions and influences, may save German society from repeating the horrors of the past.

The opposition of the Catholic Church to the direct destruction of human embryos is well known. In Germany, this position is shared by the Protestants. In April 1997, the Protestant Church of Germany issued a study document on Preimplantation Genetic Diagnosis and adopted a position against legalizing it. One of the reasons for adopting that position uses concepts that were current in the period of

National Socialism. This document of the Protestant Church states in part, "The creation of embryos with the aim of sorting out embryos with a genetic defect requires the ability to discriminate between worthy and unworthy life *(unlebeswertes Leben).*"[16] The expression "a life unworthy of life" is taken directly from the ideology of the National Socialists.

The Embryo Protection Act is another example of German society attempting to avoid a repeat of their past by providing protection for the most vulnerable in their midst through their political, legal, and religious institutions. Often opposition to embryonic research in the United States is portrayed as a matter of religious dogma. It is said that attributing any kind of value or worth to an embryo is a matter of sectarian religious belief. And yet in Germany, a modern secular state, there is a convergence of the insights of science, medicine, and religion that has given rise to a law protecting the embryo from the time of its inception.

The German state may have the support of its principal religious institutions for its strong stand in defense of innocent embryonic human life, but such a position is proving very difficult to maintain. There is pressure from researchers within Germany who want to engage in embryo research. And there is tremendous pressure from without, from others in the European Union, who want to see Germany weaken the Embryo Protection Law. In fact, in July 2004 the European Court of Human Rights in Strasbourg ruled that embryos were not human persons.

Nonetheless, up until now, post-Holocaust Germany has resisted these attempts to have any relaxation of the legal protections afforded the human embryo and other vulnerable life. The convictions of the country's religious institutions, both Protestant and Catholic, have been the same on the need to protect the human embryo. The president of the Council of the Protestant Church in Germany, Bishop Wolfgang Huber, and Cardinal Lehmann of the Catholic Bishops' Conference called on the German government in a joint statement in 2003 to urge the Council of Ministers of the European Union to adopt protective laws for the embryo along the lines of the law in Germany. The most recent formal statement of the Bishops' Conference was issued in July 2007, urging that there not be a weakening of the provisions of the Embryo Protection Law.

The religious institutions in Germany, both Protestant and Catholic, have been strong defenders of the rights of the weakest and most vulnerable in society, whether by resisting attempts to legalize embryonic stem cell research or by resisting attempts to weaken the restrictions on abortion and laws against physician-assisted suicide. As said at the outset of this chapter, medicine, science, and religion have been powerful forces in society, most particularly as they are placed in the service of those who are weakest and most vulnerable. One can only hope that this coalition can prevail in our contemporary society, which is far too prone to being driven by a utilitarian ethic, a society in which care for the sick and dying is often referred to as the "health care industry" with ideologies—be they laissez-faire capitalist or utopian—driving decisions that are not always respectful of the dignity of the human person. We must strive to see that medicine, science, and religion work together for the common good.

NOTES

1. John Paul II, *Evangelium vitae* (March 25, 1995).
2. See Stefen Kühl's *The Nazi Connection—Eugenics, American Racism and German National Socialism* (New York: Oxford University Press, 1994).
3. E. Ernst, "Commentary: The Third Reich—German Physicians between Resistance and Participation," *International Journal of Epidemiology* 30, no.1 (February 2001): 37–42.
4. Quoted in Ian Kershaw, *Hitler 1936–1945: Nemesis* (New York: W.W. Norton & Company, 2001), 255.
5. "Einführungsvortrag zu den Kursen in Alt-Rehse," Deutsches Ärzteblatt 92 (1935), 563.
6. Ernst, "The Third Reich," 37.
7. Romans 2: 14, 15.
8. Robert N. Proctor, *Racial Hygiene: Medicine under the Nazis* (Cambridge, MA: Harvard University Press, 1988), 177.
9. Kershaw, *Hitler 1936–1945*, 256.
10. Ibid.
11. Bishop Clemens August Graf von Galen, Sermon of August 3, 1941, in *Four Sermons in Defiance of the Nazis, Preached during 1941* (Münster, 1996), http://kirchensite.de/downloads/Aktuelles/Predigt_Galen_Englisch.pdf.
12. Kershaw, *Hitler 1936–1945*, 428.
13. Johannes Meyer-Lindenberg, "The Holocaust and German Psychiatry," *British Journal of Psychiatry* 159 (July 1991), 10.
14. Johannes Rau, "Will Everything Turn Out Well for Progress Befitting Humanity?"(Address of German Federal President in Berlin State Library, May 18, 2001), http://www.bundespraesident.de/en/dokumente/-,5.42675/Rede/dokument.htm.
15. Ibid.
16. Evangelical Church in Germany, "How Much Should We Know? On: Preimplantation Genetic Diagnosis," 2005, http://www.ekd.de/temp/5467.html.

CHAPTER 15

Why Science and Religion Need to Cooperate to Prevent a Recurrence of the Holocaust

IRVING GREENBERG

The Failure and the Need to Prevent a Recurrence

The Holocaust—the state-sponsored systematic attempted murder of all Jews for the crime of being and the actual mass killing of six million Jews—is in itself so horrifying and so evil that it is a natural human response to proclaim: Never Again! If people experience the humanity and reality of the victim, they intuitively commit to prevent a recurrence. Nevertheless, since 1945 there have been mass murders and attempted genocide in Cambodia, Bosnia, Rwanda, and Darfur, Sudan.[1] This is a reminder and a warning that mechanisms and specific policies must be developed if we are to successfully prevent a repetition. However, to develop effective protection mechanisms, we must first analyze what systematic failures occurred.

The successful crime reflects breakdowns on many fronts:

1. **The Failure of the German Democratic System.** Weakened by economic depression, political radicalism, and right-wing violence, the Weimar Republic was taken over by a mass movement that did not have to undergo a legitimation and filtering process to strip it of its original, ideological, murderous politics or to remove its extremist politicians. Once Nazism came to power, the democracy itself was suppressed.
2. **The Failure of the Allies to Stop Hitler's Assault.** In 1935, when Hitler took over the Rhineland, the Allies had the right under the Treaty of Versailles to remove Hitler. But the nations acquiesced to his repudiation of the limits on Germany and to the expansion of his rule. In 1938, when Hitler determined

to annex Czechoslovakia, residual war weariness and a parallel left-wing pacifism led to a similar decision not to check aggression. The European countries also refused to allow Jews at risk the extent of asylum that they required. One can cite the totally inadequate response of the nations at the Evians Conference when they were asked to provide places for Jewish refugees. Similarly, the British Government White Paper in 1938 closed the doors of Palestine for the bulk of European Jews.

3. **Christian Bystanding.** Having set up Jews as objects of hatred and dehumanization with its teaching of contempt tradition for over 1,900 years, organized Christianity was guilty of bystanding during the Shoah. Both Catholic and Protestant churches failed to outspokenly oppose the Nazis' plan.
4. **World Jewry's Failure.** Overwhelmingly, the victims failed to understand what was happening until it was too late to take evasive action or to create any force of resistance able to check the onslaught. Of course, Jews had the least margin to do anything. Furthermore, the Jews in the United States did not protest sufficiently to change American policy.

Each one of these failure factors could recur. The ability of the established democracies in Europe and America (let alone that of the newer, less stable republics worldwide) to withstand a serious extended economic depression is not beyond question. The United Nations' capacity to intervene in genocidal situations is limited and partial. Massive efforts are needed to shift policy considerations and policy constraints in all these areas if we are to assure preventive action in the future.

A good example that it is possible to change relationships and responsibilities is found in one notable achievement: the post-Shoah Jewish-Christian dialogue to rethink theology. Since the 1960s, mainstream Christian churches have come to repudiate the accusation of deicide against Jews (although it is rooted in the New Testament). They have, by and large, given up the claim of supercessionism. Some have acknowledged past failures as in the Vatican document "We Remember: Reflections on the Shoah."[2] (This Vatican confession of failures is outdone by declarations issued by the Catholic Episcopates of France, the United States, and Germany.)

The Infrastructure that Sustained the Holocaust

Since we seek a global approach to prevention of future genocide, we must go beyond the question of what failures enabled the Nazis to pursue their evil purpose to the bitter end. We must ask what systemic factors made it possible to carry on the Holocaust. The painful answer is that some of the fundamental mechanisms of modernity were essential to this genocide. That is why the Shoah could only have taken place in modern civilization and within Western culture. What were these factors?

Technology

Technology is the pride of modern culture. Technology and productivity have become a source of affluence and longer living, but, in the Holocaust, they were mobilized for murder. The Nazis organized the transportation of millions of victims to their death—especially the train transports of Western Jews (Holland, France, Belgium, Italy, etc.) across Europe in wartime to be murdered in the killing camps located in Poland. This was an immense technological feat. The Nazis also developed the technology of mass killings.

Bureaucracy

Bureaucracy is one of the great enabler institutions of modern culture. Bureaucratic protocols offer efficient, dependable procedures with equal access and decisions on the merits. The bureaucratic professional ethic offers the ability to deal with millions, and all the time. Raul Hilberg's research in particular showed that the bureaucratic mind was essential to the success of the destruction policy over a period of years. In the past, killing outbursts against Jews grew out of hatred—as in the Crusades of the Middle Ages. Paradoxically, this irrational driving force limited the planning and reduced the ability to reach the victims. Moreover, as the hatred was released, the aggressors could not keep up the slaughter campaign indefinitely.

Ideology

Ideological utopianism has developed enormous persuasive power and motivating force in modern culture. There has been so much improvement of levels of living in modernity. Due to these very breakthroughs, visions of perfecting and totally transforming the world have become credible. The utopian vision has become so persuasive and so total that it motivates people to commit great crimes—to murder people for the sake of perfecting the world. Totalitarian systems like Stalinism, Maoism, and National Socialism gave such élan to the vision of societal transformation as to enable people to overcome their natural human hesitation to murder and to dispose of millions of people.

The Ethos of Science and Scientism

Scientific discipline includes the principle of affective neutrality in doing scientific work. People are trained to separate off moral judgment in order to achieve greater methodological accuracy and objectivity in method. Such training paves the way for an instrumental ethic that equates efficiency and dependability to a moral ethic and plays down the moral effect and impact of the activity.[3] This tendency to cloak murder in the mantle of scientific ethos was made worse by the Nazis' ability to invoke eugenic and Darwinian scientific language for their policies.

Universalism

In the modern cultural ideal, rules and regulations are universal in spirit and scope. In a democracy, this translates into equal rights to participate in processes and services with no discriminatory exceptions. This honors such values as equality in a democracy in which all citizens are legally and philosophically equal. But, applied to a policy of mass killing, the principle of universal scope results in a policy in which all Jews must die. There are to be no exceptions.

THE AUTHORITY AND CREDIBILITY OF MODERNITY

The unquestioning worship of modernity even blinded the victims to what was being done to them. Alexander Donat, in his memoir of life in the Warsaw ghetto, described the torment of those who survived the initial wave of deportations to Treblinka. Why had they not anticipated the expulsions? Why had they not resisted in some organized way? After weighing various factors, Donat concluded,

> The basic factor in the ghetto's lack of preparation for armed resistance was psychological; we did not at first believe the resettlement operation to be what in fact it was, a systematic slaughter of the entire Jewish population. For generations East European Jews had looked to Berlin as the symbol of law, order and [modern] culture. We could not now believe that the Third Reich was a government of gangsters embarked on a program of genocide "to solve the Jewish problem in Europe." We fell victim to our faith in mankind, our belief that [in modern culture] humanity had set limits to the degradation and persecution on one's fellow man.[4]

UNIFICATION OF SOCIETY: GLEICHSCHALTUNG

The ultimate Nazi breakthrough that enabled carrying out the Final Solution was the policy of *Gleichschaltung,* the ability to take over all societal forces and then harness all institutions and power centers in Germany to Nazi policies. The Nazis' success in promulgating their twisted moral ethic was made possible by their ability to monopolize all alternative channels of values. One example was in the successful takeover of the German Protestant Church, including nominating pro-Nazi bishops while intimidating and silencing the Catholic Church. The Nazis conscripted all the media for this task of unification, underwriting Nazi propaganda newspapers while silencing or eliminating independent newspapers. The institutions of civil society—such as trade unions, chambers of commerce, and social fellowship groups (also potential sources of counter values)—were also taken over and neutralized.

The institutions of education and teaching were similarly co-opted. The universities were purged of Jews and of anti-Nazis. Jews were expelled from hospitals and other scientific teaching institutions and from the professional associations. They were eliminated first, but so were the anti-Nazis. In the end, these institutions actively supported (or stood by making no objections) as the Nazi policies—both anti-Semitic and antidemocratic—were enacted and enforced.

The key role that the concentration of power played in carrying out the Holocaust summons up the work of R.J. Rummel. Rummel's studies of genocide show that of the ten leading mass murders of the twentieth century, none were committed by a democracy.[5] It is not that democracies are morally pure and without the presence of evil individuals or groups that could seek to carry out destructive and even murderous policies. However, the structure of democracy incorporates the constant presence of checks and balances that break up power. In summary, modern civilization and culture have generated enormous concentrations of power. The ideology and the intention was that the power be used for life—to improve living conditions, extend longevity, increase education and culture, and expand human dignity and rights. Indeed, much has been accomplished along those lines.

Nevertheless, after the Holocaust, we understand that all this power can be brought to bear for death—for total control of society, for concentration of the victims, for waging all out war, and for mass annihilation.

It becomes essential to establish controls and limits on the exercise of power. Of these, the most efficacious limit is to pluralize the sources, channels, and centers of power so they can properly channel and limit the exercise of power. We need to break up concentrations of power while enabling the proper use of power within limits.

PLURALISM AND NEVER AGAIN

Although the Holocaust was brought to an end by the allied defeat of Nazi Germany, the forces that enabled the Shoah have continued to grow in strength and presence everywhere in the world. The military and industrial power available to aggressors today dwarf that available to Hitler. Similarly, the cultural and value trends previously described have continued unabated. Admittedly, it took the intensity of anti-Semitic hatred and the heritage of demonization of the Jews to be the spark that set off a program for total destruction of all Jews. Thus, the Holocaust remains a sui generis assault for the moment. But these trends of power and culture lend themselves also to genocide and mass murder, even if they do not translate yet into a call for total obliteration of the others. The campaign of murder and assault is unchecked in Darfur, Sudan, in 2009. Furthermore, there has been a growing revival of anti-Semitism in Europe and in the Arab countries of the Middle East. The ingredients for the marriage of ideology and technology with religious intensity and totalitarianism that could lead to a new attempt to carry out the Holocaust are in place. Certainly, one can envisage a leader like Ahmadinejad trying to make good on his threats to wipe Israel off the map if Iran obtains nuclear weapons. Under these circumstances, "never again" cannot be taken for granted. The threat of a recurrence of the Shoah is as real as is the fact that genocide has happened repeatedly since 1945.

Is there any way that we can make "never again" into more than an emotional, maybe glib, rallying cry? The processes sustaining the expansion of technology and of bureaucracy are self-reinforcing and unlikely to be stopped by countermovement organized by respondents to the Shoah. Moreover, despite the decline of the dominant ideologies of the twentieth century (Communism, Maoism, National Socialism, and various forms of authoritarian Socialism), new totalizing faiths have arisen. Most prominently, religion-based ideologies that seek perfection (as they define it) in this world, or the next, have spread formidably. They share the same capacity to seize and concentrate central control over all aspects of society, as well as the willingness to use unlimited force, "justified" by the absolute values in whose name they operate. In sum, the explosive mix of power and ideology that enables genocide is very much present in our contemporary world. This calls for a serious effort to check this ever-present danger.

The single most powerful antidote to possible repetition of Holocaust or genocide in modern society would be the articulation and promulgation of pluralism in all its forms. Many of the forms of power and central control cannot be abolished, but they can be broken up and limited, which would neutralize their potential to inflict oppression and death dealing. Since a significant element of ability to use force lies in the ideological justification of power and control, pluralism in ideas and value systems is an essential part of the mix. We need to break not only political and military concentrated power, but also to generate moral, cultural, and religious pluralism.

Political and military structures do not operate in a vacuum. Structural pluralization of power must be backed by a wide-ranging cultural and moral pluralism. Absolute authority claims pave the way for absolute applications of power to excess and, sometimes, for evil. Implicit in pluralism are limits on the authority claims of religious, moral systems and causes. In turn, those limits reduce the amount of power that can be exercised by any one group or in the name of any one worldview. Reduction of ultimacy claims paves the way for the practice of multiple religious and value systems, side by side, with none being justified in attempts to impose its will or to suppress the others.

Let it be clear that the goal is not to prevent the exercise of power by reducing all moral and value claims to level of relativism and, thereby, preventing use of power in the name of any cause. While the goal of eliminating totalizing philosophies and worldviews is justified, the moral cost of treating all views as equally lacking an authority is too high. In fact, relativism weakens the capacity of individuals and systems to oppose totalitarian systems. Few people are prepared to stand up and risk life itself on an issue where there are neither ultimate nor authority claims. Franklin Littell has written on this paradox. Nazism spoke the language of absolutes and carried on a totalitarian government and total genocide; the university and other mediating groups that might have challenged this policy of genocide were weakened in their resistance by the growth of relativism in academic circles. The debilitation affected personal responsibility and not just intellectual capability. Why give up an academic position—or risk one's life—if in the end, no cause has absolute validity?[6]

Rather, the goal must be to disseminate the philosophy and practice of pluralism—a principled commitment to absolute values, matched by an affirmation of the limits of that absolute. The recognition that absolute values are at stake evokes principled ethical stands and willingness to stand up against evil; the affirmation of limits, however, protects against aggression on the authority of one's own values. In the absence of alternate positions, absolute claims may be extended without limit to the point where they go too far or where they may metastasize, threatening other legitimate positions. In the absence of checks, even a valid belief system easily goes out of control. Mass murder and genocidal policies need the "cover" of ideological total claims to rationalize such behaviors or to overcome individual conscience and institutional checks. To the extent that absolutes are reduced by limits and acknowledgment of alternate positions, their ability to validate killings is significantly reduced. Furthermore, the acknowledgment of the other places a limit on the claims of one's own system, which paves the way to see flaws and make needed criticisms and/or corrections in one's own. In the absence of alternatives, one's position creeps into absolute status. This prevents honest criticism, correction of existing errors as well as adequate balancing of the needs and legitimate positions of the other elements in society. The pluralist welcomes the existence of alternatives (even of contradicting positions) in the belief that checks and balances will keep one's own position within its healthy limits.

SCIENCE AND RELIGION TODAY

In American society today, there are two great forces that shape the moral consensus and significantly determine what is legitimate social policy—these two forces are science and religion. Some may argue against grouping either all sciences or all religions into one category, especially since there are different approaches within the sciences and within the religions, but the fact is they can be. Each force has large constituencies, active professionals in the tens of thousands, cumulative traditions, and an inherited standing and respect. We need an articulated policy of cultural pluralism to balance and put limits on each.

Science needs to be checked with moral judgments in recognition of how its instrumental ethical and moral process may go wrong when applied to social problems. Religion needs to be checked so as not to become a vehicle to justify violence and not to be a source of demonizing of others or their alternative policies. Cultural pluralism involves acceptance of limits and affirmation of the essential value of balances and compromise. Even good scientific value systems and healthy, valid religion need these checks to stay on course.

In the pluralistic perspective, even if the religious believe that they have absolute authority and absolute revelation, and, even if scientists are convinced that science is the only reliable source of information and proven values, they would allow for the position of the other worldviews notwithstanding their contradictory teachings. This grows out of the recognition of the existence of multiple truths and of the frailty and limit of human judgment. Since each group recognizes that self-interest may corrupt one's own judgment, they accept the principle that no one group should have absolute authority. Furthermore, in light of their dignity, the others have the right to be heard even if they are wrong. Therefore, we need dialogue and reconciliation of each other's truths to proceed. All these factors justify limits on any one position or power.

Science and religion in America should work together to develop such a cooperative pluralism. There should not be one uniform source of moral consensus. If more pupils go to public school, it is good that there are private and parochial schools to offer ethical and cultural balance. In public policy, where there are culture war issues such as abortion, one should welcome the existence of clashing ethics rather than see one side or the other as bad. It is good that there is support for abortion to protect the mother's life and to uphold the right to use medicine to improve the quality of life. If there is a total ban on abortion, there is the risk of achieving ritual/ideological purity and giving that consistency a priority over the lives of people. It is also good that there is opposition to abortion. The opposition keeps reminding everyone that human life is special and precious. The opposition assures that this method does not become routine. (In countries where there is only limited access to contraception, there are higher rates of abortion, and this has led to a marked cheapening of life.)

In a society where religion is too dominant, science and scientific development are set back and restricted. This accounts for the relative weakness of science in Islamic countries currently, despite past Moslem leadership in that field in the Middle Ages. Religious tests for scientists, glorification of tradition, and repression of science that challenges tradition are marked features of religion-dominated societies.

Where science values and secular worldviews are substantially dominant (as occurred in secular societies under Marxism), religion and artistic development are retarded. In the former Soviet Union, freedom was restricted, religion was repressed and persecuted, and there was no civil society to check legal abuses or even industrial excesses (such as pollution).

Pluralism, especially one built on balance of power, is dynamic. It can, in an ongoing way, protect freedom, human dignity, and the power of free inquiry. Pluralism brings the promise of preventing the concentration of power (political, military, economic, cultural, and religious). Today, the sheer volume and overwhelming amounts of power make it of the highest urgency to build those checks into society and culture.

If American science and religion can learn the lessons of the Holocaust, they can lead the way for a dynamic model of a society so pluralist in all its institutional structures, enabled and strengthened by the presence of pluralist religions and secular systems, that there is truly a higher probability of preventing totalitarian systems or radical policies from coming to power. This is not an absolute guarantee, but it is our most promising way of assuring that Holocaust or genocide (or policies highly destructive of other groups) will "never again" be pursued.

NOTES

1. See the reflections on this repetition by Raul Hilberg, *The Destruction of the European Jews*, 3rd ed., vol. 3 (New Haven: Yale University Press, 2003), 1289–1296.
2. Holy See's Commission for Religious Relations with the Jews, "We Remember: A Reflection on the Shoah," March 16, 1998, http://www.vatican.va/roman_curia/pontifical_councils/chrstuni/documents/rc_pc_chrstuni_doc_16031998_shoah_en.html; and The International Jewish Committee on Interreligious Consultations, response to the Vatican statement, http://www.bc.edu/research/cjl/meta-elements/texts/cjrelations/resources/documents/jewish/response_We_Remember.html.
3. For a treatment of the doctors' role in euthanasia and in making selections in Auschwitz, see Robert J. Lifton, *The Nazi Doctors: Medical Killing and the Psychology of Genocide* (New York: Basic Books, 2000), especially 45–79, 172–179, 403–414, and his comments on doubling on 430–465.
4. Alexander Donat, *The Holocaust Kingdom: A Memoir* (New York: Holt Rinehart, 1965), 103.
5. R.J. Rummel, *Power Kills: Democracy as a Method of Non-Violence* (New Brunswick, NJ: Transaction Publishers, 1997); see especially the summary chapter, 203–211. See also Rummel's studies of the individual cases of mass murders in *Death by Government* (New Brunswick, NJ: Transaction Publishers, 1994); of the Nazi mass murders in *Democide: Nazi Genocide and Mass Murders* (New Brunswick, NJ: Tranaction Publishers, 1992); and the Communist case in *Lethal Politics: Soviet Genocide and Mass Murder since 1917* (New Brunswick, NJ: Transaction Publishers, 1996).

6. Franklin H. Littell, "The Credibility Crisis of the Modern University" (lecture, San Jose, California Conference on the Holocaust, May 5, 1979). Later published in Franklin H. Littell, ed., *A Christian Response to the Holocaust: Selected Addresses and Papers, 1952–2002* (Merion Station, PA: Merion Westfield Press International, 2003), 375–387. See also Eric Voegelin, "The German University and the Order of German Society: A Reconsideration of the Nazi Era," *The Intercollegiate Review* (Spring/Summer 1985), 19. Voegelin, himself dismissed from the University of Vienna in 1938, said that the Nazi appeal to students and faculty drew great strength from the contrast of the temporizing and relativistic university culture and the Nazis' promise of a striving for such absolute values as "moral perfection."

CHAPTER 16

THE STATUS OF THE RELATIONSHIP BETWEEN THE CITIZEN AND THE GOVERNMENT

WARD CONNERLY

From the founding of America, the relationship between the citizens and their government has been central to the definition of our nation. We were established as a nation of people who rely on certain "self-evident truths" to guide how we conduct our official business. Personal liberty, maximum individual freedom, and a God-given endowment of equality form the centerpiece of our system. These gifts from our Creator are unalienable—they cannot be detached from us as individuals—and the government must operate within a framework that acknowledges and protects the supremacy of these individual rights.

Every generation has to evaluate and redefine the relationship between its citizens and their government. Ours is no different as we confront terrorist activity and the response of our government to that activity; as we consider race, gender, and ethnic preferences; and as we deal with a host of other issues.

It is vitally important to understand that this relationship is never a constant or something that is etched in stone. Events and demands of our times always intrude to force the relationship in one direction or another but, generally, in favor of expanded power to the government. Rarely is the citizen on the receiving end of expanded rights and personal freedom.

Sometimes the government benefits from expanded authority as a result of the direct consent of the governed. More often than not, however, that "consent" comes by quiet acquiescence rather than by direct consent. Sometimes, the relationship is altered in subtle ways, on an incremental basis. For example, assume that an event with significant influence on society occurs and that legislation is enacted to address that specific event. The Civil Rights Movement of the 1960s comes to mind. Initially, the 1964 Civil Rights Act was enacted to extend to black Americans their right to equal treatment. In this instance, the rights of other citizens were unaffected by the granting of equal rights to black people.

The relationship between private citizens and the government was also unaffected by this extension of basic rights to blacks. Over time, however, additional legislation and policies were approved by the government to further benefit blacks and other "minorities" as well as women. With each new enactment, not only were the rights of non-blacks and males reduced, but also the power of the government was increased to make decisions about such matters as "diversity." Moreover, as matters evolved, the question of the relationship between the private citizen and government became secondary to the more specific issue of race and gender preferences and their effect on respective groups. Yet, it is undeniable that the basic citizen-government relationship has been reordered by the incremental initiatives of the government.

At other times, an event of substantial consequence occurs and cataclysmic reforms are put in place to address such an event. We need look no further than September 11, 2001, to find such an event. In response to this event, more Draconian security measures were deemed essential to protect the people. With the passage and enforcement of such measures, a major restructuring of the relationship between private citizens and the government was instantly, substantially, and perhaps, irrevocably achieved.

While few can deny the legitimacy of the rationale for greater governmental power to protect the security of our nation, the result is an unmistakable expansion of the government and a severe diminution in the rights of the citizen. Witness the intrusive, although understandable, inspections of persons and property at airports.

The great challenge of a democracy is how to successfully reconcile the issues that threaten to diminish the role of the citizen and those that would have the effect of embellishing that of the government. Unfortunately, this challenge cannot be answered as long as the American public is unaware of the threats that are presented by certain government policies and actions. This unawareness often opens the door to that "slippery slope" that is sometimes feared in public policy.

In a recent visit to the Holocaust Museum Houston, I was moved profoundly by the exhibits depicting atrocities committed against human beings because they were of the "wrong race." One would think that this experience would be sufficiently compelling to forever sour Jews about the use of "race" by the government to favor some citizens over others, especially when one considers that quotas were often used to deny admission to Jews to some American universities. Yet, public opinion polls reveal majority support among Jews for race-based preferences.

As I viewed those exhibits, I could not help but think, how can Jews who support race-based affirmative action not understand the fine line between racial genocide and racial benevolence—both of which flow from the same premise that the human family is divided into separate "races," some of which are good and some of which are bad? Not only has the wrong answer to this question been responsible for a radical departure from the constitutional guarantee of equal protection of the law for every American citizen, but it has been responsible for so many other problems throughout our history.

This thought also prompted me to reflect upon my own personal history as an American. At the time of my birth, I was classified as a "colored" person—a fact that is confirmed by a "c" on my birth certificate. Although my ancestors were of African, American Indian, Irish, and French descent, according to the odious "one-drop" rule, nothing mattered except the African portion of my complex mix of ancestries. Even today, as more and more Americans marry across these lines of "race," my own grandchildren, who are all of me and their Irish grandmother, are classified as "black," despite the fact that two of my four grandchildren have a mother who is half-Vietnamese. It angers me that my government still subdivides the American people into what amounts to separate and distinct "food groups"—Asian, black, Latino, Native American, and white. And, to think that once upon a time, the government enforced laws prohibiting individuals from marrying outside their "race." It should be troubling to us all when the government considers it appropriate to classify and distinguish among its citizens on the basis of "racial" characteristics.

In 1994 James Cook, the highly qualified grandson of poor white sharecroppers and laborers, was denied admission to the University of California (UC) at San Diego Medical School. His parents prepared a report, entitled the Cook Report, which demonstrated that in one year's pool of medical school applicants, there were 50 who were better qualified academically than the best affirmative action applicant admitted to UC San Diego Medical School. Yet, none of these applicants was admitted to *any* medical school in the United States, which all operated under governmental affirmative action legislation and regulations.

The Cook example underscored the egregious extent to which race was being used by UC medical schools and other components of the UC complex. Not only was this use of race in violation of the Supreme Court ruling that race could be used as "one of many factors," it eroded the quality of the institution and represented a gross injustice to applicants who were denied admission because of their race, gender, or ethnic background.

Fortunately, there are 23 states in America where the people have reserved for themselves the right of self-governance through the citizen initiative process. Already, four of those 23—California, Michigan, Nebraska, and Washington—have determined that the issue of racial distinctions among its citizens by the government is of sufficient importance to the relationship between citizens and the government that those states forbid government agencies from making such distinctions in the areas of public contracting, public education, and public employment.

The principle of "equal treatment" is at the core of American democracy. The guarantee of equal rights for all Americans is not something to be bargained away by racial groups or to be misappropriated by the president or the Congress or any state legislature. It is the American Constitution, which is derived from the founding principle that equality is a gift endowed by the Creator, which informs the relationship between the citizen and the government.

While the status of the relationship between the citizen and the government is a muddled one at present, as long as the people understand that the principle of equal rights is a primary means by which the government seeks to expand and exercise its control over citizens to achieve certain government-desired objectives, and as long as the citizens do all within their power to resist altering that principle of equal rights in America, then there is hope that private citizens can retain control over their government.

CHAPTER 17

FROM NUREMBERG TO THE HUMAN GENOME: THE RIGHTS OF HUMAN RESEARCH PARTICIPANTS

HENRY T. GREELY

In 1946 and 1947 a military court of the United States, as the occupying power for Southern Germany, tried 20 physicians and three nonphysician Nazi officials for crimes committed in the course of human experiments performed in German concentration and prisoner of war camps. Seven of the defendants were acquitted; nine were sentenced to prison terms of varying lengths, and seven were sentenced to death. On June 2, 1948, the seven condemned men were hanged by the neck until dead.[1]

The Nazi Doctors' Trial, *United States v. Karl Brandt et al.*,[2] led to the promulgation of the Nuremberg Code for conduct of human experimentation, but it did not end abuses of people participating in research. Along with subsequent scandals, it has largely ended the physical mistreatment of research subjects. Today's issues are both less horrible and more subtle, but are nonetheless important in how we treat the people who give, quite literally, of themselves for biomedical research.

In this chapter, I will examine some ethical difficulties in one form of modern biomedical research—the large-scale genomic database and repository (hereafter "genomic biobanks"). I start with a brief review of the development of our system for protecting participants in human research. I will follow that with a discussion of the nature and growth of these biobanks. I will then analyze three different ethical and legal challenges posed by this kind of research. I hope to convince you that we need to rethink the ethics of such research in ways that will better protect not just the safety, but also the broader interests of research participants.

OUR SYSTEM OF HUMAN SUBJECTS PROTECTION: ITS HISTORY AND CURRENT FUNCTIONING

The current rules and ethical thinking underlying the protection of participants in human research are, of course, shaped by the history that produced them. This history has led to a useful, if occasionally frustrating, set of procedures and rules, but procedures and rules that overlook some important interests. I will set out the history that has produced the current American regime for human subjects research and then describe how that system now works.

HISTORY

Short of the Big Bang, it is rare that the history of anything has a logically compelling starting point, and the protection of human research subjects, in the United States and elsewhere, is no exception.[3] Although I could choose several earlier episodes,[4] the Nazi Doctors' Trial is probably the least bad starting place.

The war crimes and crimes against humanity for which some of the defendants were convicted include many kinds of revolting actions that had nothing to do with human experiments, from the murder of concentration camp inmates to the murder of hundreds of thousands of mentally ill or retarded Germans (and others). But the aspects of the crimes most relevant here are the human experiments conducted on unwilling concentration camp subjects or prisoners of war. These experiments included seeing how long it took subjects to freeze to death, how long people could survive drinking sea water, the effects of low atmospheric pressure, the use of radiation to sterilize subjects, and the value of various medical interventions for diseases or conditions that were intentionally inflicted on subjects (malaria, septic wounds, wounds from mustard gas, and burns from phosphorus bombs).[5]

The defendants argued, among other things, that no international law governed human subjects research. This argument led Dr. Leo Alexander, a physician who was consulting with the prosecution, to prepare a memorandum for the tribunal on appropriate human experimentation. The tribunal reworked this memorandum into ten points that were included in its judgment and later referred to as "The Nuremberg Code." The first point began with the unequivocal statement, "The voluntary consent of the human subject is absolutely essential." This first point, on informed consent, is the longest and most detailed, but other points include the need to have a good scientific reason for the research, one sufficient to "justify" the performance of the experiment, and the importance of limiting both risk and any harm to the subjects. The Nuremberg Code was not legislation and had no legal power beyond the jurisdiction of the military tribunal (i.e., the U.S. sector of occupied Germany), but it had great importance as the first widely known discussion of the ethics of human subjects research.

The next important event in the history of human subjects protections came not from a court but from the World Medical Association (WMA), an organization made up of national medical societies, such as the American Medical Association. In 1952 the Association created a permanent Medical Ethics Committee, which in 1953 took up the idea of creating a position paper on research involving human subjects. A draft was completed and introduced at the 1961 meeting of the WMA's General Assembly but was not finally adopted until the 1964 meeting in Helsinki. Because of its source, the declaration is expressly aimed at laying out the ethical duties of physicians, not of other researchers. The WMA has revised its Helsinki Declaration six times, most recently in 2008 in Seoul.[6]

While the Nuremberg Code and the Declaration of Helsinki were widely known, no legal authority applied either to research in the United States. In 1966 it became clear that American research was not always following the principles of either document. Henry Beecher, an anesthesiologist at Harvard Medical School, published an article in the *New England Journal of Medicine* that detailed 22 examples of unethical research in the United States.[7] Also in 1966, but four months *before* the Beecher article, the United States Surgeon General published a policy requiring all institutions that received research funds from the Public Health Service (which includes the National Institutes of Health) to provide prior review of human subjects research. This policy was supplemented in 1971 by an "Institutional Guide to DHEW Policy on Protection of Human Subjects."[8]

In 1972 these issues reached a much broader public with the public exposure of the so-called Tuskegee study.[9] This research project, undertaken by the U.S. Public Health Service with government and philanthropic funds, had a local base at the Tuskegee Institute.[10] At its start in 1932, the study enrolled about 600 men, mainly poor sharecroppers. About two-thirds were already infected with syphilis; the other third was to serve as a control group. The men were not told of the nature of the study, that they had syphilis, or anything about the disease. Local physicians were told not to treat them. Until the end of World War II, the only treatments for syphilis were relatively risky and ineffective, as well as expensive. After the war, however, penicillin became widely available as a safe, inexpensive, and effective treatment for most cases of syphilis. The Public Health Service considered whether to terminate the study and provide penicillin, but decided instead to continue the observational study as, in light of the new treatments, it would be the last chance to obtain this kind of data.

The Tuskegee study had never been secret, and articles based on it were openly published; yet it avoided much public attention until July 1972 when a newspaper account exploded around the country. The study was reviewed within the government and, several months later, terminated. By that time, 28 of the men had died from syphilis and roughly another 100 died from conditions that could be complications of syphilis.

The revelations about Tuskegee, together with other scandals, led to substantial congressional interest, notably through hearings chaired by Senator Edward Kennedy. As a result, in 1974 Congress passed the National Research Act. Senator Kennedy had been pushing for a national commission, independent of the National Institutes of Health (NIH), to oversee the protection of human subjects; others resisted an external reviewer. Ultimately, the Department of Health, Education and Welfare turned its earlier administrative policies into binding regulations, and the legislation was changed to transform the proposed regulatory commission into the purely advisory National Commission for Protection of Human Subjects of Biomedical Research and Behavioral Research.[11]

This prolific commission published many volumes of discussion of ethical issues in research, but its most important contribution may have been its shortest, the so-called Belmont Report. This report, officially titled "Ethical Principles and Guidelines for the Protection of Human Subjects of Research," summarized the commission's conclusions about the relevant ethical principles and their application to human subjects research.[12] The commission identified respect for persons, beneficence, and justice as the most relevant principles and saw informed consent, assessment of risks and benefits, and selection of research subjects as each principle's most important application. The report was published in the *Federal Register* on April 18, 1979.

Meanwhile, the newly renamed Department of Health and Human Services (HHS) continued to wrestle with regulations for human subjects research. Influenced in part by the National Commission, it published draft regulations in August 1979, which were finalized, with substantial changes, in January 1981. The issue of how to regulate human subjects research was then taken up by another commission, the President's Commission for the Study of Ethical Problems in Medicine and Biomedical and Behavioral Research. It advocated broader application of the HHS regulations.[13] Following this advice, the federal government embarked on interagency negotiations that resulted, in 1991, in 16 federal agencies adopting a form of the HHS regulations as their own regulations, regulations that, as they were common to all these agencies, became known as the Common Rule.

THE COMMON RULE

So what *is* this Common Rule? It is the main legal regulation of human subjects research in the United States. It applies to human subjects research performed by the federal agencies that have adopted it (which include almost all agencies with any substantial research funding), or research funded by those agencies, *or* research performed by an institution that has given the federal government (typically HHS) an assurance that all of its relevant research will conform to the Common Rule. The Food and Drug Administration's (FDA) version of the Common Rule also applies to research done with FDA authorization or with an intention to submit the results to the FDA. The "institutional assurances" coverage is probably the most important element, as every major research university, college, or institute has given such an assurance, making all of its relevant research covered. "Human subjects research" includes both research that requires a new intervention (from surgery to an interview) with a living person and research that uses "identifiable private information" about a living person, even if someone else has collected it.

Apart from a few exclusions and exemptions, not relevant here, any human subjects research done by a covered institution must follow the Common Rule. That largely means that the research must be approved through a process that involves an "Institutional Review Board," universally referred to as an IRB. IRB decisions must be recorded, but they are not normally made public, and, unlike a judicial opinion, they typically do not require any justification. Each institution's IRBs sit alone; they are not part of any coordinated regional or national system. HHS does maintain an Office for Human Research Protections, which provides guidance and regulatory oversight with respect to the Common Rule, but it does not review particular IRB decisions.

In theory the Common Rule provides for an expert body to scrutinize all human subjects research to make sure it adequately protects the people involved. The real world looks a little different, at least at major research institutions.

The IRB process imposes substantial costs in time and effort, from researchers and from IRB members and staff. It is not clear how well always overworked IRBs can actually assess research protocols. It often seems to researchers that the time spent in the IRB process far outweighs any conceivable risks, particularly when the research involves no physical intervention in the subject. The process also leads to long and complex consent forms. In spite of the Common Rule's command that the form be "in language understandable to the subject," 10-, 15-, or even 20-page consent forms are not rare. Given the evidence that subjects rarely understand or remember the terms of the consent form, as well as the uncertainty about the extent to which the actual research really conforms to the IRB's commands, it is fair to ask whether the Common Rule makes sense.

The answer, I think, must be yes. In spite of all of its flaws, the Common Rule serves as a protection against much risky and inappropriate research. Even apart from the value of the actual review, the mere fact that researchers have to specify and justify their use of research participants is important. As a result, human research subjects have largely, if not entirely, been protected from direct risks to their safety. Although the process is, no doubt, often inefficient and susceptible to streamlining, it has protected human subjects from some kinds of risks. It has not, however, provided protection for all of the relevant interests of possible subjects, a reality brought into high relief by the rise of genomic biobanks.

GENOMIC BIOBANKS: A NEW KIND OF RESEARCH

Three new technologies have made possible new kinds of human subjects research, different in both degree and kind from what went before. The technologies are the computerized infrastructure that allows the storage and manipulation of vast quantities of data; new statistical methods for analyzing that data; and new methods for *ascertaining* the relevant data. This is true throughout modern biomedical research but perhaps nowhere as true as in genomics,[14] the study of large quantities of data of the nucleotide base sequence of deoxyribonucleic acid (DNA). Genomics is now giving rise to genomic "biobanks," which, by collecting DNA samples and health (and other) information from individuals, allow researchers to compare sequence data and health conditions from many people at once. These biobanks, as we will see, in turn give rise to new ethical—and legal—issues about the rights of the donors of that DNA and health information.

THE RISE OF LARGE-SCALE GENOMIC BIOBANKS

Genomic biobanks are based on the hope that big numbers can make up for small risks. If we cannot spot a genetic variation that increases risk by 10 percent by looking at several hundred or thousand people in higher-risk families, maybe we can find it by looking at 10,000 or 100,000 or 1,000,000 people. This is the basic justification for large-scale genomic biobanks.

Biobanks combine DNA samples from people (and sometimes other kinds of physical samples, such as blood samples or tumor biopsies) with lots of information about their health, their environment, and other possibly relevant traits. Of course, traditional, small family-based genetic research involves biobanks as well, although that term is usually not used. But the new approach focuses on quantity—of subjects and of information. Researchers can analyze now more than a million specific locations, known as single nucleotide polymorphisms (SNPs), using a single SNP chip for less than $500 a person. And the long-heralded world of electronic medical records is finally arriving, so the biobank may not have to collect and enter health data, but just copy it from a clinical database.

One could create a large-scale databank to study one particular disease, but this does not make economic sense. The cost of analyzing an entire genome is the same whether you use it for one disease or every possible disease. So, more or less, is the cost of transferring health data, particularly as even in studying one disease, a researcher cannot be sure whether health data about other diseases could be important. No one will want to spend a billion dollars for a biobank solely to study non-insulin-dependent diabetes when, for a few dollars more, the biobank could be used to study *every* disease.

And so, the new genomic biobanks will be not only large in scale, but large in scope. Instead of aiming at one particular disease, they can be used to study any disease; instead of belonging to (or, at least, being created and used mainly by) one researcher or one small team of researchers, they are created by thousands of people for potential use by tens of thousands of researchers, all over the world. The UK Biobank, the product of a collaboration between the British National Health Service and the Wellcome Trust, a (wealthy) private foundation, will be the first, but it may not be the last.

Genomic Biobanks and the Rights of the People who Participate in Them

The change from traditional, family-based genetic research to large-scale genomic biobanks has implications for the people who take part in the research by contributing their DNA and their health information. In the old style, research subjects enrolled in genetic research because that research focused on a disease common in their family. It offered hope, if not for them, then for their children and grandchildren. And they participated because a researcher dedicated to learning about, and ultimately treating, *that* disease had built a relationship with them.

Large-scale genomic biobanks are not about one family's disease. Your sample may be used for research on a disease common in your family or it may not. And it does not involve a personal connection with a particular researcher. Your DNA will be collected by one of hundreds of staffers working as part of a massive project. Your health information may be obtained, in part, by someone's interview with you, but mainly by a download from a clinical medical database. If the DNA sample is also obtained from a medical sample, you may have no personal interaction with anyone from the research project.

These differences change, fundamentally, the real relationship between the research and the research participants. Those changes raise a set of new ethical and legal challenges. I have written about many such challenges in other contexts; in this chapter, I will describe only three: the problems of consent, the problems of confidentiality, and the problems of return of significant personal information. In each case, genomic biobanks risk creating research participants who feel—with good reason—cheated and bitter.[15]

THE DISAPPEARANCE OF REAL CONSENT

The Nuremberg Code, the Helsinki Declaration, and the Common Rule all require that research participants, with only a few exceptions, have given their informed consent to take part in the research. Under the Common Rule, they need to be told the purpose of the research, the procedures to be followed, and the plausible risks and benefits to themselves, in understandable language. The goal, according to the Common Rule, is for the information to allow them to make a fair and free decision whether to participate.

Traditional genetics research could do that. The researcher could tell the potential participants that she was looking for genes linked to one particular disease. The potential benefits would be more knowledge about the disease, with the possibilities of genetic testing, prevention, or effective treatment. The main risks were the psychological and familial risk of knowing that you were, or were not, at high risk for that particular disease, as well as the possibility of societal discrimination based on that risk.

But what can you tell people about their participation in a genomic biobank? The purpose is to look at anything that any researcher in the world thinks might have a genetic association, from diseases to non-disease characteristics to human evolution and history. The processes to be followed include taking a DNA sample, analyzing it, and putting it in a database with health information—but what after that? The benefits might affect any possible disease—or possible non-disease trait. The risks are not limited to knowledge—your own, your family's, or society's—of your risk for one disease, a disease likely already common in your family, but the knowledge of your risks for potentially any genetic disease or condition. And with that data—both the genomic information taken from your DNA and the data from your health information—available to any or almost any researcher anywhere in the world, you have to worry about its confidentiality in the hands not just of one researcher, but of tens of thousands of researchers you have never met.

It is, I believe, impossible for anyone to give truly informed consent to participation in a large-scale genomic database. The information needed to make that consent informed just does not exist, as the purpose, the processes, the benefits, and the risks will be different from continent to continent and from decade to decade. One could give a kind of blanket permission, but that permission is *not* informed consent.

The NIH disagrees. Although acknowledging some of the problems of blanket consent, the NIH Office of Human Research Protection (OHRP), which provides guidance to the IRBs applying the Common Rule, allows it.

In a way, it has no choice. To outlaw blanket consent is to prohibit any broad, multiuse resource using human data or samples. This gives up a great deal of scientific potential, all for, arguably, a purely speculative concern. After all, why shouldn't we let competent adults decide whether they are willing to take the risks inherent in genomic biobanks?

In fact, the concerns are not purely speculative. One very specific risk is that the research participant's DNA and health information will be used for purposes she did not foresee and with which she strongly disagrees. At one extreme, one could imagine a DNA sample—or, in the future, the whole DNA sequence even without an existing sample—being used to create, without subsequent consent, that person's clone. Unless the consent form for such a large-scale biobank explicitly forbids this use, there seems nothing legally to prevent it.

More plausibly, the person's data or sample could be used for research she hates. Personally, I am extremely skeptical about any research on race, genetics, and intelligence, particularly if conducted by any "researcher" with access to the Internet and then the biobank. I would be extremely angry if my DNA or data were used in such research, but I would have given blanket consent.

This example is not purely hypothetical. A Native American tribe called the Havasupai live in a side branch of the Grand Canyon. About 300 Havasupai live on the reservation with another several hundred tribal members scattered around the country. In the late 1980s, a researcher from Arizona State University approached the Havasupai to do a genetic study of type 2 diabetes, which hits their tribal members at devastatingly high rates. The tribal government, and many individual members of the tribe, agreed to the research, but only for studying diabetes. More than a decade later, the tribe learned, by chance, that the researchers had not only used their samples and data to study the genetics of diabetes but also to study the genetics of schizophrenia, as well as the possible history of the tribe.

The tribe and its members were outraged. They had not wanted, or consented to, this kind of research, either about a stigmatized mental illness or about their own history. Rightly or wrongly, they believe that their people were created where they now live and that any scientific support for a different origin, particularly the Bering Strait hypothesis, sometimes abbreviated by some Indian people as "the BS hypothesis," would hurt their ongoing political and legal disputes about land. Lawsuits against the University and various researchers followed—lawsuits that after several years remain unresolved. And future genetics researchers are unlikely to be welcomed by the Havasupai.

But what is to be done?

One possible step would be to be bluntly honest with new consent forms. Potential research subjects could be told that their materials and data could be used for any purpose at all, even one they did not like, without any notice to them or opportunity to object. That would not make up for the absence of informed consent, but it would at least increase our confidence that the participants knew what they were getting into. It would also be fought tooth and nail by researchers, who would fear that it would cut down the number or people willing to volunteer as research participants.

One could imagine going back to participants in biobanks to ask their (now informed) consent for each and every proposed use of their samples and data. This raises enormous logistical problems; keeping track of research subjects, even over the space of a year or two, can be very difficult. But if the researchers do keep track of all the participants, how will participants react to being asked for their informed consent time after time, for arcane projects—once a year, once a month, once a week, or, for very successful biobanks, once a day? Participants are likely to find that kind of full informed consent more a curse than a blessing.

There are more realistic alternatives. One could ask research participants, when giving informed consent, to describe research in which they do not want to participate. Or, from the other direction, ask them to list the kinds of research they are willing to be part of. Rather than relying on the participants' thoughts at that time, the consent form could include checklists of possible research subjects with boxes for "yes" or "no."

Even this method is ungainly. Neither the participants on their own nor the researchers who create the checklists will be able to foresee all the possible uses or to understand all the possible implications of that research. A final possibility is to make this kind of consent-like process continual, but not at the individual level. A committee of research participants, as proxies for the whole group of people in the genomic biobank, could review all specific research proposals. This committee could decide whether the research would likely be controversial or uncontroversial with research subjects. If they found it was too controversial, they could refuse to let it go forward. If they found it might be controversial with some substantial part of the biobank participants, they might let it go forward only with new, individual "real" informed consent from the participants—those who were happy with it could allow it; those who were not could opt out. (Another, less powerful, version of this approach would inform participants of all newly proposed research projects that want to use the biobank and allow the participants individually to opt out, or, depending on where the presumption is set, to opt in.)

Several solutions have been proposed but none is perfect. Each solution tries to preserve some kind of consent and control by research participants over the research that is done with their DNA and health information. Otherwise, we ignore their interests in how their materials are used, at the risk of prompting a powerful backlash. We need to require some kind of process that makes informed consent for genomic databases more real—or, at least, less empty.

The NIH is not, however, moving in this direction. Quite the contrary. Its proposal to accumulate biobank data does not require anyone at NIH even to decide that the original consent included language about broad future uses. A research subject who gave consent for research on type 2 diabetes by a form that included some vague language about "and other uses" might find that his DNA is being used to study genetic influences on sexual orientation, mental illness, drug addiction, race, or anything else. And that is wrong.

THE IMPOSSIBILITY OF GUARANTEEING CONFIDENTIALITY

Consent forms for research usually talk about maintaining the confidentiality of the participants' records and information. Sometimes they speak of "ensuring" confidentiality; other times they merely state that the researchers will "do their best" to keep the information confidential. They never tell the participants just how uncertain their confidentiality, in fact, must be.

In a world of computerized data, confidentially *cannot* honestly be guaranteed. Every year there are scores, if not hundreds, of major security breaches of confidential information in the United States, usually of people's employee or financial records. Some of those come from clever and devious computer hackers, but most of them are because someone loses or has stolen a laptop or a flash drive. Biobank information is equally susceptible to loss or theft, particularly if researchers all over the world can access them. Even if downloading the data is prohibited, it may well happen and then uncounted laptops and flash drives have to be protected.

Of course, if the data does not include anything that identifies an individual subject, this may not matter. And, typically, researchers use coded identifiers for individuals. Henry T. Greely becomes 28Q5da16. If that improperly accessed database contains only the code, how can the participant be hurt? In several ways.

First, these coded systems contain keys, documents or files that say that "Henry T. Greely = 28Q4da16." Otherwise, the researchers could never add any new information to the file. Even if they got new health data or genetic data on me, they would not know what file to put it in. Those keys can themselves be hacked, lost, or stolen. The keys can also be abused, if someone with legitimate access to the key uses it for illegitimate purposes. The more people with access, the greater that risk. Celebrities can never count on their hospital records remaining confidential as long as hospital employees with access to the computer system are interested. There are ways, such as audit trails, to reduce these risks, but not to eliminate them.

But let's say there is *no* key—the information has been anonymized and all connection has been lost and forgotten between "Henry T. Greely" and "28Q4da16," even at the cost of giving up the possibility of continuing data entry. Confidentiality is still not assured. Any rich database will contain information that others might be able to use to identify.

This is easy with genomic data. It takes very little genomic information to identify any one person (at least, any person who does not have a living identical twin). As genomic data becomes more common—in doctors' offices, in genetic genealogy companies, in criminal databases, in paternity testing centers—it becomes easier and easier for someone with access to your genomic information from any of those sources to identify you in the biobank database . . . and, therefore, to get access to all your health and other information in that database.

But even the health or other information can be identifying. I am a white male, born on June 25, 1952, in Columbus, Franklin County, Ohio. The Franklin County birth records (probably soon to be available online thanks to the strong demand from genealogists for data) would name the children born in Columbus that day. There were probably about eight of us: four female, four male; six white, two black. There is a good chance that just looking at those birth records, for a medium-sized county, would allow you to narrow down, dramatically, the identity of anyone in a biobank file containing date and place of birth. It then becomes fairly easy to look for facts about me in the biobank to distinguish me from the others born on that day. My height, my weight, my hair color or eye color could all help identify; so could my health history.

"Re-identification" of "anonymous" subjects could be accomplished through their genetic information, their health information, or their demographic information. Several years ago, LaTanya Sweeney, now a computer science professor at Carnegie Mellon University, demonstrated that quite dramatically. William Weld, governor of Massachusetts at the time, had made all the health records of Massachusetts state employees or Medicaid recipients available for research, but with all "identifiers" removed. Within a couple of days, Sweeney had Weld's health records delivered to his office. She had re-identified him based on the health records, the voter registration records, and driver's license records.

Of course, one could limit the information in these research files. Instead of recording date of birth, one could record year of birth; instead of place of birth, state of birth. The problem is that each time data is made fuzzy for identity purposes, some scientific value may be lost. We know, for example, that both season of birth and location of birth can correlate with the risks of different diseases. Those may be important variables to include in the genomic biobanks, and it is impossible to know in advance what will or will not turn out to be important.

It would seem, then, that people considering participating in a genomic biobank should at least be told that the confidentiality of their data and samples cannot be guaranteed. The chance of having their identity leak out—through hackers, lost or stolen equipment, or other problems—is not small. The chance that a determined someone could identify them from their data in the biobanks is substantial, and it grows every day as more genetic, demographic, or health information becomes available to people online.

No one requires such frank and honest information about confidentiality today. And that is wrong.

RETURN OF IMPORTANT INFORMATION

Sometimes research is part of providing clinical medical care, as when a surgeon tries a new technique or an internist uses an experimental drug on a patient. Research in genomic biobanks, however, will rarely be part of medical practice. Instead, researchers who have never met the participants (who may or may not be "patients," because they may or may not be under a physician's care) will mine the information in the biobank to try to learn something about genetics and diseases or traits. But what happens if the researchers learn something about a participant that is medically important?

Right now, often nothing. Long ago, the Office for Protection from Research Risks, the predecessor to OHRP, issued guidance that told IRBs that researchers should have three choices about providing information about individual risks to research subjects. They could say they would provide such information; they could say they had the discretion to provide such information; or they could say that they would not, under any circumstances, provide that kind of information. Researchers have overwhelmingly chosen the third option. They have probably been encouraged to do so by their institutions' lawyers, who may reason that by disclaiming any intention to provide this kind of information, the institution cannot be sued successfully for not providing it.

There is a certain legal logic to this position, but it has the disadvantage of, among other things, being profoundly immoral. Consider this situation. A genomic biobank analyzes DNA from 100,000 research participants. It finds that about 50 of them have mutations in genes called MLH1 or MSH2, which put those participants at very high risk of a kind of colon cancer known as hereditary non-polyposis colon cancer (HNPCC). If the participants knew of their risk, they could manage it, through regular screening colonoscopies or, in some cases, by the prophylactic removal of the colon. Either of those options is better than metastatic colon cancer, which kills about 50,000 Americans each year.

The biobank and the researchers who use it now *know* that these participants are at great risk of death, a risk they could greatly lessen by providing timely information. The biobanks and its researchers are benefiting, scientifically and (at least in terms of grant funding) financially, from the contributions participants have made to the biobanks. How can they refuse to take steps to save the lives of these same participants? And, yet, researchers have not generally accepted the idea that they have a duty to provide that kind of potentially life-saving information. Of course, providing such information is not simple.

More importantly, it is hard to know *when* to provide this kind of information, in two respects: how much do we have to know about the genetic risk, and what kind of risk does it have to be? In traditional research into human genetic disease, the researchers will be looking for a link between a disease and a particular genetic variation. If they are still in the process of discovering, understanding, and proving the nature of this kind of link, they may not have solid information to give participants. Just how certain these researchers, or, perhaps, those in the field more generally, have to be about the genetic link is an important and difficult question. This question—of disclosing results that are still uncertain—probably underlies much of the opposition to returning results. But biobanks are in a different situation. Instead of trying to discover a new link, the starting material for biobanks—the SNP analyses or, soon, the full-length sequences of the entirety of the participant's genome—will already include information about well-known and well-established risks. This issue will remain for associations that are newly uncovered by research with the biobanks, but not for older risks.

In terms of the kinds of risk that warrant disclosure, HNPCC is an easy case. Almost everyone with the mutation will get the disease; the disease is often deadly, but there are steps that can greatly minimize its risks. How high a risk does a genetic variation have to confer before the participants need to be informed: 50 percent? 10 percent? 1 percent? How serious does the disease have to be: highly life threatening, somewhat life threatening, disabling, or just very unpleasant? Does it matter whether there is an effective, well-proven treatment, a speculative treatment, or no treatment at all? There is no treatment for Huntington's disease, but we know that a substantial minority of those at risk for it choose to be tested for it, sometimes to help them in planning their lives (including decisions about parenting), sometimes just to know. Should biobanks have to return information about that kind of disease? Should the biobank have to notify research participants about risks not to themselves but to children they have, or may have? Research participants will invariably be carriers for several nasty recessive diseases, not affected by them personally but, if they have children with another carrier, at risk for having children with the disease.

It is also hard to know *how* to deliver this information. We would not want a biobank to call up a participant and leave a voice mail: "Remember that DNA sample you gave for research? Well, we analyzed it and found out that you are going to get a deadly disease." Should information be provided only in person, or are telephones, e-mails, or the mail appropriate? Should it be provided only by a genetic counselor or a clinical geneticist? Or should the information be provided directly to the participant's personal physician, who can then tell the participant? What if the participant does not have a physician? What if he does not have health coverage at all?

Some participants will not want to know about disease risks. How do we honor their wishes? It is hard to image a conversation that begins, "I am calling from the biobank; we have learned something about you that we think you may want to know. Do you want me to tell you about it?" At the same time, if participants are asked in advance, in a general way, whether they want to know about genetic disease risks uncovered in the research, their answer may not have been made with the actual situation in mind. A person who says "no" might be thinking about Huntington's disease, for example, rather than relatively preventable HNPCC.

Whom to notify is another question. The research participants may be strongly affected by the information, but, depending on the nature of the genetic condition, the participants' parents, siblings, and children may also be at substantial risk. Should the biobank notify them? Should it tell the participants that they should notify them? Two courts in the United States have given opposite answers to just those questions in lawsuits against physicians caring for patients with genetic diseases.

Finally, there are pesky logistical questions. People change their homes, their phone numbers, and their e-mail addresses. How hard should a biobank have to try to find an at-risk research participant? Medical knowledge also changes. What if, five years after a participant's DNA has been added to the biobank and her genome sequence has been analyzed, a strong new association is proven between particular genetic variations and a serious disease? Does the biobank have an obligation to keep going back to the participants' data to see if they have any risks that should be disclosed? If so, how often and for how long?

How to go about providing research participants with significant information discovered about them as a result of the research raises issues that are both real and vexing. Research institutions have good reasons to fear getting involved in this potential quagmire. But, in reality, they will have no choice. This is not just a moral issue; it is a practical one.

In spite of possible disclaimers of any intent to provide this kind of information, one could make a legal case that the institution has a duty to provide it. Physicians, psychotherapists, and other health professionals clearly have an obligation to warn their patients of health risks. These kinds of obligation also apply to treating physicians while engaged in research with their patients. Courts might well find, in a case with compelling facts, that even researchers who were not treating physicians had a duty to research participants, probably through holding that they were in a "fiduciary" position. Alternatively, courts might find that the research participants reasonably expected to be told of such results, even in the face of language in the consent form that said otherwise, especially as the consent forms get longer and longer. Some recent research indicates that most people assume researchers would provide information on such risks and are surprised to learn that it is not automatic.

Even though the legal theories for holding the biobank and its sponsor liable for nondisclosure are not, at this point, certain winners, with the right facts such a case could be very attractive to a plaintiff's lawyer. Imagine a case where the grieving widow, with her young children, sues over the death of her husband from HNPCC when she can prove that the biobank had, easily accessible in its database, information that he was at risk. Even if the biobank's legal case had some merit, its public relations position would be terrible. The universities, foundations, or government agencies that own biobanks cannot want the reputation that they do not care about the people who participate in its research. "Go to the University Hospital and become a guinea pig" is not a message they will want broadcast.

But the issue goes beyond individual biobanks and their supporting institutions. It goes to the heart of biomedical research. I can think of nothing more likely to embitter and enrage research participants as the researcher disregarding clear, serious, and preventable risks to those participants. Participants would feel cheated and misused—and rightly so. Their willingness to take part in further biomedical research would certainly plummet, as, most likely, would their political support for research. So, in the end, whatever its legal status, an unwillingness to provide significant information to research participants about their own risks is just wrong—morally, practically, and for the future of biomedical research.

CONCLUSION

Genomic biobanks are a very long way from Nazi concentration camps. The NIH is not the SS. Biomedical researchers are not modern Mengeles. Those differences are some evidence that we have learned from the past. The Nuremberg Code, the Helsinki Declaration, and the Common Rule *do* work. Sadistic torture masquerading as research, high risks of death or serious physical injury, and frank coercion are gone. Those are good things.

But as times change, so do the challenges they present. Modern biomedical research faces its own challenges. The fact that they are much more subtle than those raised during the Holocaust does not mean they are unimportant.

People whose DNA, tissue samples, and health information are used in medical research never get rich as a result and rarely get cured. They have the right not to be abused, but they have other interests as well—interests in how their materials are used, in knowing the truth about confidentiality, and in getting useful personal information when researchers can provide it.

I believe in biomedical research. I do not necessarily support and cannot necessarily defend each and every research project all around the world, but in general,

I believe today's research will greatly reduce the burden of human suffering—something that is not only morally permissible but morally compelled. But that does not mean we can ignore the interests of the participants without whom research would be impossible—participants who are giving, quite literally, *of themselves* to let research proceed. They trust us with their samples and secrets; abusing their trust is just wrong.

But, for biomedical research, it is more than just wrong. It is "worse than a crime, it is a blunder." The Tuskegee study is often cited as one cause (of many) why African Americans are less likely than others to participate in biomedical research. The Havasupai are unlikely to participate in genetic research again anytime soon; other Native American tribes may also be affected. We cannot keep making these mistakes. We need to treat research participants with complete respect, acting toward them as we would ourselves like to be treated (not least because we all may, with or without our knowledge, end up as participants in genomic biobanks).

American lawyer Leon Jaworksi served as an army lawyer during and just after World War II. In that capacity, he prosecuted several war crimes trials and, just before his discharge, helped the prosecution prepare for the Nazi Doctors' Trial. Jaworski was deeply and permanently affected by what he saw and learned about the Holocaust and other Nazi atrocities, but his conclusions were not just directed to the fate of Germany. Ultimately, he concluded, "it would be a mistake for any nation smugly to feel her people immune to such a fate."[16] As we face the different, more subtle, but quite real ethical issues raised by genomic biobanks, we need to heed his advice.

NOTES

The author thanks his research assistants, Kelly Lowenberg and Mark Hernandez, for their help in this chapter.

1. For a broad discussion, see George J. Annas and Michael A. Grodin, eds., *The Nazi Doctors and the Nuremberg Code: Human Rights in Human Experimentation* (New York: Oxford University Press, 1992).
2. *Trials of War Criminals before the Nuremberg Military Tribunals under Control Council Law No. 10, Vols. 1 and 2.* (Washington, D.C.: U.S. Government Printing Office, 1949) (hereafter cited "Trials of War Criminals").
3. This is only a sketch of the history, and, although I have tried to go beyond the common six-step formula (The Nazis to Helsinki to Beecher to Tuskegee to the National Commission to the Common Rule) to give some hint of the complexity of the story, I recommend Ruth Faden and Thomas Beauchamp, *A History and Theory of Informed Consent* (New York: Oxford University Press 1986), chapters 5 and 6, 151–234.
4. Ironically, the first legal regulation of human subjects research may have been in Prussia, one of the constituent parts of Germany. After scandals involving research by a Dr. Albert Niesser, the Prussian ministry issued a directive regulating human subjects research and requiring, at least in nontherapeutic research, informed consent of the subjects. A similar directive was issued by the Health Ministry for all of Germany in 1931. See Jochen Vollmann and Rolf Winau, "Nuremberg Doctors' Trial: Informed Consent in Human Experimentation before the Nuremberg Code," *British Medical Journal* 313 (1996):1445–1447.

Another good starting point could be the yellow fever experiments conducted by Dr. Walter Reed in Cuba, on military volunteers, paid Cuban and Spanish subjects, and public health service physicians themselves, but, like the Prussian experience, it generated little lasting attention or discussion.

5. The specific experiments and actions considered by the tribunal are discussed in Trials of War Criminals, Vol. 2., 175–180.
6. The current version of the declaration can be found on the website of The World Medical Association, World Medical Association Declaration of Helsinki: Ethical Principles for Medical Research Involving Human Subjects, October 22, 2008, http://www.wma.net/e/policy/b3.htm.
7. Henry K. Beecher, "Ethics and Clinical Research," *New England Journal of Medicine* 274 (1966): 1354–1360.
8. For more discussion of these governmental actions, see Faden and Beauchamp, *A History and Theory of Informed Consent,* 206–212.
9. For a comprehensive discussion of the Tuskegee study, see James H. Jones, *Bad Blood,* 2nd ed. (New York: Free Press, 1993).
10. One of the many ironies of the Tuskegee study is that the study, completely directed by the federal government, is commonly known by the name of the African American university that only made its facilities available to the government.
11. Faden and Beauchamp, *A History and Theory of Informed Consent,* 213–214.
12. National Commission for Protection of Human Subjects of Biomedical Research and Behavioral Research, "Ethical Principles and Guidelines for the Protection of Human Subjects of Research," *Federal Register* 44, no. 23 (April 18, 1979): 23192–23197.
13. President's Commission for the Study of Ethical Problems in Medicine and Biomedical and Behavioral Research, "The Adequacy and Uniformity of Federal Rules and of Their Implementation: Protecting Human Subjects," *Federal Register* 48, no.146 (1983): 34408–34412.
14. The dividing line between (old) genetics and (new) genomics is not clear; indeed, it is not entirely clear whether "genomics" is anything more than a newer and more appealing way of saying "genetics." To the extent that the terms are used differently, "genetics" usually refers to the study of one or a few genes, either in abstract form or through their DNA sequences. "Genomics" is usually used to refer to studying large amounts of DNA sequence information and, in some cases, the entire sequence of a person's (or nonhuman organism's) genome.
15. I have made these arguments before, several times, in more detail and with more references. Those interested in reading more can find it in Henry T. Greely, "The Uneasy Ethical and Legal Underpinnings of Large-Scale Genomic Biobanks," *Annual Review of Genomics and Human Genetics* 8 (2007): 343–364; Henry T. Greely, "Informed Consent and Other Ethical Issues in Human Population Genetics," *Annual Review of Genetics* 35 (2001): 785–800; Henry T. Greely, "Human Genomics Research: New Challenges for Research Ethics," *Perspectives in Biology and Medicine* 44, no. 2 (Spring 2001): 221–229; Henry T. Greely, "Iceland's Plan for Genomics Research: Facts and Implications," *Jurimetrics* 40 (Winter 2000), 153–191 ; Henry T. Greely, "Breaking the Stalemate: A Prospective Regulatory Framework for Unforeseen Research Uses of Human Tissue Samples and Health Information," *Wake Forest L. Rev.* 34 (fall 1999), 737–766; Henry T. Greely, "The Control of Genetic Research: Involving the Groups Between," *Hous. L. Rev.* 33 (1997):1397–1430. See also North American Regional Committee, Human Genome Diversity Project, "Proposed Model Ethical Protocol for Collecting DNA Samples," *Hous. L. Rev.* 33 (1997):1431–1473.
16. Leon Jaworski, *After Fifteen Years* (Houston, TX: Gulf Publishing, Co., 1961), 13.

CHAPTER 18

MEDICAL PROFESSIONALISM: LESSONS FROM THE HOLOCAUST

JORDAN J. COHEN

INTRODUCTION

The principal objective of medical educators is, and has always been, to prepare future physicians to meet the needs and expectations of the society they are pledged to serve. Today's medical educators face many challenges in trying to meet that objective. In addition to ensuring that medical students and residents acquire the knowledge and skills required of competent clinicians, they also must deal with a host of novel issues raised by the complexities and turbulence of contemporary medical practice. Among those issues are:

- continued escalation in health care costs
- rapid advances in complex medical technologies
- explosion of scientific knowledge
- wide racial and ethnic disparities in health and health care
- large gaps in the quality of health care service
- widespread concern about patient safety
- inequities in access to health care services
- unexplained variations in patterns of practice

I argue that one challenge facing medical educators rises above all of these: how to sustain medical professionalism in the face of increasing threats from commercialism. In appreciating the urgency of surmounting this challenge, educators would be well advised to heed the lessons gleaned from the behavior of physicians during the Holocaust. I would not suggest for a moment that the modern-day threat of commercialism is any way comparable in magnitude to the unconscionable and unprecedented menace posed by Nazism. But I do suggest that the Holocaust can teach us both how vulnerable physicians are to the seductive appeal of a self-serving ideology and how devastating the consequences can be when physicians abandon the ethical precepts of their calling.

MEDICAL PROFESSIONALISM

Let's consider first what is meant by the term "medical professionalism." Medical professionalism comprises the actions individual physicians must undertake to fulfill the profession's implicit contract with society.[1] That contract calls for physicians to apply their knowledge and skills exclusively for the benefit of patients and the public, in return for which physicians and the medical profession are accorded uncommon privileges (e.g., the ability to regulate their professional affairs, a high degree of autonomy in practice, social status, and relative economic security). The actions that professionalism requires of physicians are based on a set of timeless principles and attendant responsibilities. A contemporary articulation of the principles and responsibilities of medical professionalism can be found in "A Physician Charter: Medical Professionalism in the New Millennium," drafted by a working group of the American Board of Internal Medicine Foundation, the American College of Physicians Foundation, and the European Federation of Internal Medicine.[2] The charter identified three underlying principles: the primacy of patient interest, patient autonomy, and social justice. It also identified ten categories of physician responsibilities, as shown in table 18.1.

Physicians throughout history have faced many obstacles in attempting to maintain their commitment to professionalism. Human nature alone, with its strong motivation for self-interest, can trump professionalism's appeal for altruism. And there has always been pressure to abandon one's professional responsibilities from those peers who appear to profit from unprofessional behavior. Add to that the ease with which physicians, acting in the privacy of the doctor/patient relationship, can abuse their patients' inherent vulnerability, and one can appreciate the strength of character required to adhere to the lofty tenets of professionalism.

Table 18.1 Physician Charter: Responsibilities of Individual Physicians

A physician should

- maintain professional competence throughout one's career
- deal honestly with patients
- respect patient confidentiality
- avoid inappropriate relations with patients
- avow scientific knowledge
- fulfill professional responsibilities
- improve the quality of health care
- advocate improved and equitable access to care
- support the just distribution of scarce resources
- maintain trust by managing conflicts of interest

Each of these conventional threats to professionalism continues to challenge today's physicians, but none is more worrisome, in my opinion, then the threat of commercialism.

THE THREAT OF COMMERCIALISM

Commercialism's fundamental ethic is based, not on the *subordination* of self-interest, as demanded by professionalism, but rather on its opposite, the *fulfillment* of self-interest. Some of the terms commonly employed in describing transactions governed by these contrasting "isms" are listed in table 18.2 and illustrate the different attitudes and orientations called for by each. These differences are pointedly summarized by their associated mottos: *caveat emptor* (buyer beware) in the case of commercialism and *primum non nocere* (first, do no harm), in the case of professionalism. Commercialism calls for wariness on the part of customers entering a marketplace. By contrast, professionalism calls for trust on the part of patients entering a doctor's office. Trusting that doctors will *voluntarily* place their patients' interest ahead of self-interest is professionalism's hallmark. Once lost, the likelihood of the medical profession regaining public trust is vanishingly small.

Why worry if commercialism succeeds in replacing professionalism as the underlying ethic of physicians? Which is to ask, why worry if commercialism nullifies medicine's social contract? The answer has much more to do with what patients and society would lose by way of objective guidance to better health than with what the profession of medicine would lose by way of its privileges. Today's powerful diagnostic and therapeutic modalities, for all the potential benefits they offer, also have the potential, through both omission and commission, to cause great physical and great financial harm to patients. Nothing can protect patients from these potential harms like a trustworthy physician. No laws, no federal regulations, no government agency, no amount of fine print in a health insurance contract can substitute for a trusted physician honor-bound to place patients' interest uppermost.

Table 18.2 Contrasting Terms: Professionalism and Commercialism

Professionalism	Commercialism
Doctor	Provider
Patient	Customer
Trust	Suspicion
Caring	Pandering
Services	Commodities
Values	Margins
Cures	Profits
Pride	Bonuses
Primum non nocere	*Caveat emptor*

RELEVANCE OF THE HOLOCAUST

Let's review what we know about German doctors and the Holocaust. We know that the Holocaust was initiated under the aegis of the Nazi Party's "Euthanasia Program," which aimed to exterminate all "undesirable" people (e.g., Jews, Gypsies, mentally ill, physically disabled).[3] To implement the program, doctors were required to play two essential roles. First, doctors were required to "diagnose" the individuals who were deemed to be "undesirable" and, by doing so, to sanction their extermination.[4] Second, doctors were required to identify and develop the highly efficient extermination techniques needed to accommodate the large numbers of individuals destined for extermination.[5]

We also know—and this is the most relevant fact to bear in mind—that doctors participated in the Euthanasia Program of their own *free will*, not because the Nazi government forced them to do so. There is no evidence that coercive measures were needed to enlist the cooperation of a sufficient cadre of physicians to accomplish the program's unspeakable objectives. Indeed, a sizable number of doctors volunteered to take part without even having to be asked.[6]

Why did they do so? Putting aside those relatively few who were sadistic psychopaths or worse, we know that many (arguably most) of those who carried out the Euthanasia Program were ordinary, "good" doctors.[7] Without the acquiescence and willing participation of a large part of the German medical establishment, the Holocaust simply could not have taken place. Hard as it is in retrospect to imagine how "good" doctors could have so willingly abandoned the core precepts of their profession, the facts are inescapable. They saw participation in the Euthanasia Program as dutifully complying with the prevailing ideology and, in many cases, as advancing their own self-interest professionally and economically.

According to Dr. Elise Huber, Berlin president of the German Medical Association, "Today we know and must accept the responsibility that the medical community was [involved], and that community remained silent... It was... medical megalomania that paved the way for the Nazi ideology and the Holocaust."[8]

Sherwin B. Nuland writes, "To my startled dismay, I found myself understanding why so much of the German medical establishment acted as it did. I realized that, given the circumstances, I might have done the same.... What we learn from history comes far less from the study of events than from the recognition of human motivation—and the eternal nature of human frailty."[9]

The Holocaust, then, is not to be viewed as a horrifying tale about a unique historical aberration, but as a cautionary tale for all time about what it is to be a human being.

IMPLICATIONS FOR MEDICAL EDUCATION

The failure of medicine to adhere to its ethical grounding in the face of seductive public policies is what allowed the horrific consequences of Nazism to occur. It would, of course, be ludicrous to equate the dangers of commercialism's intrusion into medicine with the total corruption of the profession as fomented by Nazism. But, the startling reality that the Holocaust teaches us is the ease with which a contemporary ideology—one that promises a better future for the country—can undermine ordinary, "good" doctors' core ethical obligation to the primacy of patients' interest. Indeed, experimental psychologists have substantiated how easily "good" people can be made to do "bad" things.[10] Recognizing how readily "good" doctors can be led to act contrary to patients' well-being underscores medical education's vital role in strengthening the resolve of future physicians to sustain their commitment to professionalism.

How can medical education help to ensure that future physicians will sustain that commitment, whether it is threatened by commercialism or by any other seductive ideology that might come along? I suggest six strategies for medical educators to adopt in addressing this important objective:

1. **Stress the importance of character as well as academic prowess as criteria for admission to medical school.**

   Given that well over 90 percent of entering medical students graduate with an M.D. degree, admission to medical school is tantamount to admission to the practice of medicine. Thus, medical school admissions committees have a critically important role to play in maintaining the medical profession's high ethical standards. The challenge admissions committees face was captured succinctly by Herman Nothnagel, the nineteenth-century German physician who described atrial fibrillation; he wrote, "Knowledge attains its ethical value and its human significance only by the humane sense in which it is employed. Only a good person can become a great physician."[11]

   Despite the difficulty of assessing such traits as honesty, integrity, compassion, and altruism, admissions committees have done a remarkably good job over the years of selecting students with the requisite attributes to become "great" physicians. In the face of today's realities, however, the selection process should pay even more attention to the "moral fiber" of medical school applicants and their natural inclinations toward serving others.

2. **Establish explicit learning objectives for professionalism, and evaluate student achievement of those objectives.**

   Professionalism is a way of *acting*. As noted above, it comprises a set of observable behaviors (see table 18.1). Medical educators should not only indicate clearly that those behaviors are the expected norms of the profession but also should implement appropriate evaluation regimes to assess individual student achievement of those behavioral norms. Formative assessments are especially valuable as a way to identify opportunities for improvement before undesirable habits become ingrained. Assessing professionalism is admittedly more difficult than assessing many other aspects of student achievement. Fortunately, educators are devoting considerable effort at developing better, more reliable methods for this purpose, and those efforts should be encouraged.[12]

3. **Nurture, recognize, and reward humanism.**

   Professionalism and humanism are intimately linked. Whereas professionalism comprises a set of behaviors, humanism comprises a set of deep-seated personal convictions about one's obligations to others, especially others in need. Thus, one can view humanism as providing the passion that animates professionalism.[13]

In the absence of a strong humanistic disposition, individual physicians might well act in such a way as to fulfill all the expectations of professionalism without actually believing in the virtues or principles that underpin them—going through the motions, so to speak. By contrast, humanistic physicians are intuitively and strongly motivated to adhere to the traditional virtues and expectations of their calling. Consequently, physicians who harbor the passion of humanism are best positioned to remain steadfast in fulfilling their professional responsibilities despite ever-present temptations to do otherwise.

Consequently, a decisive way for medical educators to reinforce their students' commitment to professionalism is by nurturing their students' humanistic qualities. A variety of means have proven effective in furthering this objective, including rituals such as the White Coat Ceremony and awards that honor exemplars of humanism among students and faculty physicians. But, by far the most effective means for achieving this goal is through "good-old-fashioned" role modeling. The true values of the profession are communicated to students and residents far better by what they witness respected faculty (and fellow students) actually doing than by what they hear in the classroom or read in textbooks.

4. **Promote an institutional culture that models professionalism: the informal curriculum.**

As just noted, the learning environments in which medical students and residents acquire their clinical skills are themselves powerful "teachers" of the profession's actual values. The lessons learned by observing how faculty and more senior learners interact with one another, with patients and their families, and with other staff constitute what has been called the "informal" curriculum.[14] Unfortunately, those learning environments are frequently crucibles of cynicism rather than cradles of professionalism. As a consequence, the wholesome attributes that the vast majority of students exhibit on entry to medical school are often suppressed rather than reinforced during the arduous period of medical education and training. If future physicians are to withstand the threats to professionalism that are sure to come, it is incumbent on medical educators to ensure that the institutional cultures to which impressionable learners are exposed model, and, thereby, inculcate the true values of the profession.

5. **Establish the rationale for adhering to the precepts of professionalism: the formal curriculum.**

Acknowledging the critical importance of the informal curriculum is not to imply that the formal curriculum has little or no role to play in transmitting the profession's values. Quite the contrary. There are a variety of ways in which lectures, case-based seminars, assigned readings and the like can contribute importantly to this task. Examples of topics that can be communicated well by the formal curriculum are: the historical development of the modern conception of professionalism in medicine and elsewhere; the contemporary definition of medical professionalism; the inherent threats to professionalism posed by human nature and peer pressure; and the vulnerability of patients and the ease with which doctors can abuse their privileged relationship with patients.

6. **Incorporate relevant humanities in the curriculum.**

   Educators should take special note of the effective use that can be made of sources drawn from the humanities to depict both professional and unprofessional conduct. Paramount among these sources are historical accounts of the role of physicians and the medical profession in enabling the Holocaust. No more powerful example exists of the overarching importance of physicians remaining steadfast in their fidelity to the tenets of professionalism. Every medical student should become fully cognizant of the disastrous consequences that stemmed from the failure of German doctors to adhere to the fundamental admonition of their calling: *primum non nocere*—first, do no harm. Fully appreciating the lessons imbedded in annals of the Holocaust would go a long way toward ensuring that future physicians sustain their commitment to medical professionalism and that the public continues to enjoy the unique protections that medicine's social contract promises.

SUMMARY

The Holocaust is unquestionably an extreme example of flagrantly unprofessional behavior on the part of physicians and the entire medical profession of the time. It cannot, however, be dismissed as an unrepeatable example of how "good" doctors were lulled into doing very bad things by a uniquely compelling ideology. Indeed, what countless doctors did during the Holocaust is now understood to be consistent with a common human predilection. This conclusion has been amply corroborated by experimental psychologists who have documented how easily ordinary, "good" people can be induced to do very "bad" things by the context in which they are placed or find themselves.

This chilling lesson from the Holocaust is highly relevant to the challenge posed by commercialism's intrusion into the workings of contemporary medicine. While no one would suggest for a moment that commercialism poses anything remotely like the threat that Nazism did, the Holocaust is a most vivid reminder that an ideology alien to professionalism can consciously or unconsciously lure physicians away from their ethical duty to uphold the primacy of patients' interest. The ethic of commercialism is alien to the ethic of professionalism and poses a real threat to the continued existence of a medical *profession* worthy of the name.

Medical educators have key roles to play in strengthening the resolve of future physicians to sustain their commitment to professionalism. They are admonished to stress the importance of character as well as academic prowess as criteria for admission to medical school; to establish explicit learning objectives for professionalism; to evaluate student achievement of those objectives; to nurture and reward humanism; to recognize the importance of the institutional culture (the "informal" curriculum) in transmitting the values of professionalism; to use the formal curriculum to establish the rationale for adhering to the precepts of professionalism; and to incorporate in the curriculum relevant humanities, especially historical accounts of the roles played by physicians and the medical profession in the Holocaust.

NOTES

1. Sylvia Cruess, "Professionalism and Medicine's Social Contract with Society," *Clinical Orthpaedics and Related Research* 449 (2006): 170–176.
2. ABIM Foundation, ACP-ASIM Foundation, and European Federation of Internal Medicine, "Medical Professionalism in the New Millennium: A Physician Charter," *Annuals of Internal Medicine* 136 (2002): 243–246; and Medical Professionalism Project, "Medical professionalism in the new millennium: a physicians' charter," *The Lancet* 359 (2002): 530–532.
3. Walter Laqueur, ed., *The Holocaust Encyclopedia* (New Haven, CT: Yale University Press, 2001).
4. Sue Fishkoff, "They Called It Mercy Killing," *The Jerusalem Post*, April 12, 1996.
5. Laqueur, *The Holocaust Encyclopedia.*
6. Ibid.
7. Daniel J. Goldhagen, *Hitler's Willing Executioners: Ordinary Germans and the Holocaust* (London: Little Brown, 1996).
8. Fishkoff, "They Called It Mercy Killing."
9. Sherwin B. Nuland, review of *Deadly Medicine: Creating the Master Race Exhibition* (U.S. Holocaust Memorial Museum), *The New Republic* (September 2004).
10. Stanley Milgram, *Obedience to Authority: An Experimental View* (New York: Harper Collins, 1974); and James Hollis, *Why Good People Do Bad Things: Understanding Our Darker Selves* (New York: Gotham Books, 2007).
11. Who Named It?, "Carl Wilhelm Hermann Nothnagel," http://www.whonamedit.com/doctor.cfm/2735.html.
12. David Thomas Stern, ed., *Measuring Medical Professionalism* (New York: Oxford University Press, 2006); and Richard Cruess et al., eds., "Professionalism Mini-Evaluation Exercise: A Preliminary Investigation," *Academic Medicine: Journal of the Association of American Medical Colleges* 81, 10 supp (October 2006): S74–S78.
13. Jordan J. Cohen, "Linking Professionalism to Humanism: What It Means, Why It Matters," *Academic Medicine: Journal of the Association of American Medical Colleges* 82, no. 11(November 2007): 1029–1032.
14. David Thomas Stern, "In Search of the Informal Curriculum: When and Where Professional Values Are Taught," *Academic Medicine: Journal of the Association of American Medical Colleges* 73, 10 supp (October 1998): S28–S30.

CHAPTER 19

ASSESSING RISK IN PATIENT CARE

GEORGE PAUL NOON

As practicing cardiovascular surgeons, we often treat seriously ill and high-risk patients. I was an associate of Michael Ellis DeBakey in a developing field of cardiovascular surgery. We were referred patients who could not be cared for or operated on elsewhere. They came for life- and limb-saving operations. Dr. DeBakey's goal was to provide these patients with optimal care. We were willing to undertake new high-risk procedures in patients as long as there was a potential for recovery. For many of these innovative procedures, operative risk and long-term outcome were not yet determined. Our risk assessment was developed by clinical experience. With experience, risk assessment improved. Today, risk assessment has expanded beyond the patient.

Risk assessment of a patient starts with a general medical evaluation of the patient and the type of treatment anticipated. These two factors will initiate the determination of the risk of complications and outcome of the treatment. Other factors that influence risk assessment include the patient's location at the time of assessment (rural or urban community versus tertiary hospital), associated illness, available treatment options, social and psychological factors, and the financial status.

When assessing a patient's risk for major adverse events and outcomes during or after a surgical procedure, the clinician depends first on clinical evaluation and judgment. The risk assessment may be further refined through additional invasive and noninvasive testing.

For assessment of patients who undergo thoracic surgical procedures, including heart and/or lung transplantation, at The Methodist Hospital, Houston, Texas, a variety of algorithms and strategies are used to assist in categorizing patients' risk-benefit ratio. Numerous algorithms have been developed that are helpful in the risk assessment and subsequent treatment decision. Some of these are Acute Physiology And Chronic Health Evaluation (APACHE II) severity of disease classification;[1] The Society of Thoracic Surgery (STS) risk calculator;[2] the International Society for Heart and Lung Transplantation (ISHLT) risk-factor analysis;[3] United Network for Organ

Sharing (UNOS), which provides a database of transplant procedure and outcomes;[4] Scientific Registry for Transplant Recipients (SRTR);[5] and the Methodist Hospital Medical Review Board (MRB).

So, for example, STS offers outcome analysis in the areas of adult and congenital cardiac surgery and general thoracic surgery, with the objective of improving the patient's quality of care and outcomes of cardiothoracic surgery. For this purpose, STS has created a national database on surgical procedures and has developed a risk calculator for thoracic operations, which considers the patient's demographics; previous hospitalizations; risk factors (smoking, diabetes); previous operations; cardiac status (New York Heart Association, or NYHA, classification); medications; hemodynamics; and the planned procedure in its risk calculation. The health care staff is then provided with information about the risk, anticipated morbidity and mortality, length of stay, risk of stroke, prolonged ventilation, infections, renal failure, and re-operation. This information can help the physician plan necessary intervention and anticipate complications, which are related to the overall preoperative health of the patient.

Assessing the risk of patients undergoing surgical procedures and making the decision for surgery, however, remain processes that are individually tailored to each patient. The decision is made on a case-by-case basis and considers the risk of potential adverse outcomes (death) versus potential benefits, including substantial improvement in quality of life.

Once the risk assessment has been completed, the final decision for the patient's care will be made. Patients who are suitable medical candidates may be rejected because of other factors in their evaluations, for example, inadequate financial resources before and after therapy, history of noncompliance, lack of adequate care at home, and drug addiction. High-risk patients may be rejected or referred elsewhere, if possible, since anticipated outcomes and length of stay could cause loss of insurance contracts and federal funding. The patient, his family, or medical power-of-attorney and associates may also decline treatment for personal reasons. The ethics committee may be called upon to help make final decisions and recommendations.

On this note, I will briefly present the well-known case of my mentor, Dr. Michael E. DeBakey, who on December 31, 2005, experienced sudden onset of severe pain in the right side of the neck, which was initially evaluated and treated at home. A few days later, on January 3, 2006, his suspicion of a possible dissection of the thoracic aorta was confirmed by magnetic resonance angiography (MRA). His diagnosis was DeBakey-Type II dissecting aneurysm. He initially chose medical therapy and observation. The aneurysm measured 5.7 centimeters in diameter and had an intramural hematoma without evidence of an intimal flap. Despite his medical condition, on January 6, 2006, Dr. DeBakey chose to lecture to the Academy of Medicine, Engineering and Science of Texas on "Mechanical Circulatory Assistance." His presentation was superb. A lecture following his was "Aortic Aneurysms and Dissections: Using Genetics to Unravel Pathways to These Fatal Diseases." He remained at home until January 23, 2006, when his medical condition worsened, mandating hospitalization. Serial imaging studies revealed a progression in aneurysmal size, which correlated with a deteriorating overall clinical status. Dr. DeBakey chose at that time to continue his medical therapy, however. On January 29, 2006, he indicated, when questioned, that in case of a cardiopulmonary arrest, he did not want to be resuscitated. At no point during the course of his illness, did Dr. DeBakey unequivocally refuse surgery when speaking with me. He told me that if he were unable to make decisions for himself, we should proceed "to do what needed to be done."

On February 9, 2006, an MRA demonstrated enlargement of the aneurysm to 7.7 centimeters, now with an intimal flap, increased pericardial fluid, and bilateral pleural effusions. These changes were significant and suggested bleeding from the aneurysm, with a high risk for fatal rupture. The patient was sedated with Versed and Ativan for the MRA diagnostic imaging procedure. His clinical status and therapeutic options were discussed with the family. Dr. DeBakey was not able to participate because of sedation and his general condition. The family, including his wife, daughter, three sons, and two sisters decided on surgery, knowing the risks and alternatives. His wife signed an operative permit, and we scheduled the operation. The anesthesia staff decided not to participate in the operation because of the patient's age (97 years), the risk, and the no-code status in the chart for cardiac arrest. The anesthesiologists discussed their position with Dr. DeBakey's family. Prior to this illness, Dr. DeBakey had been very active, traveling nationally and internationally. The risk of surgery was very high, but we thought, justifiable. Dr. DeBakey's family made arrangements for an outside anesthesiologist, who had previously worked with Dr. DeBakey, to provide the anesthesia. At this point, a hospital ethics committee convened to discuss the decision for the proposed operation on Dr. DeBakey, and a final decision was made to proceed with the operation.

Dr. DeBakey underwent an operative procedure that DeBakey and Associates developed. The type II dissecting aneurysm of the ascending aorta was resected, replaced with a Dacron graft (developed by Dr. DeBakey) with resuspension of the aortic valve and the use of profound hypothermia and circulatory arrest with total cardiopulmonary bypass and retrograde cerebral perfusion.

Dr. DeBakey had a prolonged recovery requiring temporary hemodialysis and tracheostomy for respiratory support. With the aid of his physicians, nurses, physical therapists, and the support of his family, he managed to recover from his illness and return to work. Even though Dr. DeBakey had a difficult and prolonged recovery, he stated he was grateful we proceeded with his life-saving operation.

Postoperatively, Dr. DeBakey faithfully engaged in physical therapy, gained strength, and returned to his office, where he continued to be consulted by colleagues and officials for advice on a wide variety of subjects. He participated in a number of official events, including the groundbreaking for the Michael E. DeBakey Library and Museum at Baylor College of Medicine and traveled to New York to attend a Lasker Board meeting and to Washington, D.C., to receive the prestigious Congressional Gold Medal from President George W. Bush on April 23, 2008.

On Friday afternoon, July 11, 2008, Dr. Antonio Gotto, a long-time colleague and friend, visited Dr. DeBakey at home. They discussed preparing a new edition of their best-selling book, *The New Living Heart,* and Dr. DeBakey was enthusiastic about collaborating on it. In the late afternoon on that day, I also visited Dr. DeBakey, had gumbo with peppers from his garden and a delightful conversation with him. He was looking forward to continuing his physical therapy and his professional schedule in the coming week. Approximately one-and-a-half hours later, I received a call that

Dr. DeBakey "had collapsed" suddenly at home and was being brought by ambulance to the hospital. I was shocked, having seen him so shortly before. He died unexpectedly that evening, 58 days before his hundredth birthday. An era of his spectacular surgical innovations and his inspiring leadership in medical education, research, and patient care came to an unwelcome close, but his legacy of excellence and dedication to the improvement of humanity will live on forever as all who knew him mourn his passing. It is an especially severe loss to me personally, but I am thankful for the long and inspiring personal and professional relationship I had with this international medical icon.

NOTES

1. Acute Physiology and Chronic Health Evaluation (APACHE II), http://www.sfar.org/scores2/apache22.html.
2. The Society of Thoracic Surgeons (STS), http://www.sts.org/; and The Society of Thoracic Surgeons, Risk Calculator, http://209.220.160.181/STSWebRiskCalc261/de.aspx.
3. The International Society for Heart and Lung Transplantation (ISHLT), http://www.ishlt.org/.
4. United Network for Organ Sharing (UNOS), http://www.unos.org/; United Network for Organ Sharing, Lung Allocation Score (LAS) calculator, http:www.unos.org/resources/frm LAS Calculator.asp?index=98; The United Network for Organ Sharing, MELD/PELD calculator, http://www.unos.org/resources/meldPeldCalculator.asp; and The United Network for Organ Sharing, CPRA calculator, http://www.unos.org/resources/frm CPRA Calculator.asp.
5. Scientific Registry for Transplant Recipients, http://www.ustransplant.org/.

CHAPTER 20

JEWISH MEDICAL ETHICS AND RISKY TREATMENTS

AVRAHAM STEINBERG

JEWISH ETHICAL PRINCIPLES

The basis, validity, and source of Jewish ethics are rooted in the belief in God and His Torah (Bible), whereas the basis of secular ethics is primarily humanism and rational intellect.

The following are some basic principles of Jewish ethics as viewed by Orthodox Judaism:

- In Judaism, there is no basic difference between laws/regulations and morals/ethics because both are integral parts of the Torah, and their validity flows from the power of the Torah and the Divine revelation. Therefore, basic principles, discussions, and debates on Jewish *ethical* issues do not differ from those of Jewish *legal* issues.
- Jewish ethics includes the guidelines for proper conduct for man in relation to his fellow man as well as in man's relation to God. Therefore, there is no difference in the binding nature of the law between the prohibitions of stealing, killing, falsehood, revenge, carrying a grudge and the like, and the laws prohibiting idol worship, Sabbath desecration, eating on Yom Kippur, and the like. So too, there is no difference between the obligations of giving charity, visiting the sick, burying the dead, caring for orphans and widows, and the like and the observance of dietary laws—such as eating unleavened bread (*matzah*) on Passover—and sitting in the *Sukkah* on Tabernacles (the holiday of *Sukkot*), and the like.
- According to the Torah and Jewish law, one is obligated not only to refrain from doing bad, but one must do good by being compassionate and charitable with one's fellow human beings as it is written, "Turn from evil and do good."[1] These are two equal parts of the Jewish ethical obligation. Therefore, not only are harmful acts such as stealing, wounding, and killing prohibited, but there

exist positive commandments: to give charity, to visit the sick, to be hospitable, to return lost objects, and the like.

- These Jewish principles require not only proper acts but also proper thoughts and intentions. The Torah forbids hatred, covetousness, revenge, and carrying a grudge and requires one to love God, to love one's fellow man, and to love a stranger, in spite of the obvious difficulties in controlling one's thoughts.

The Bible and Talmud are replete with references to proper conduct, both between man and man and between man and God. Jewish ethical teaching involves general concepts and principles on the one hand and specific rules and regulations on the other. The Bible cites a number of basic principles about the proper relationship between man and man, such as:

- "Love your fellow man as yourself"[2]— "This is a major principle in the Torah";[3] "What is hateful to you, do not do to your neighbor, that is the whole Torah, while the rest is commentary, go and learn it."[4]
- "Do not profane the name of your God."[5] Namely do not conduct yourself in a way that profanes the name of God.[6]
- "You shall do what is righteous and good in the eyes of the Lord."[7]
- "Observe justice and perform righteousness."[8]
- "Despise evil and love good, and establish justice by the gate."[9]
- "Do justice, love kindness, and walk humbly with your God."[10]
- "Thou shall not stand aside while your fellow's blood is shed."[11]
- "Righteousness, righteousness shall you pursue."[12]
- "Fill the earth and subdue it."[13]
- "That you may walk in the way of the good, and keep the paths of the righteous."[14]
- "Its ways are ways of pleasantness, and all its pathways are peace."[15]

However, the Jewish ethical system, like the Jewish legal (Halakhic) system, is not satisfied with general theoretical rules alone; it is filled with practical and individual guidelines. The Torah requires every human being to strive for perfection in one's conduct vis-à-vis another person in actions, in speech, and in thought—not just abstract general good behavior.[16]

Jewish Medical Ethics Principles

In Jewish medical ethics, there are obligations for both physicians and patients.

- There is an obligation upon the physician to heal the sick. The role of a physician is not optional in Jewish law; it is obligatory.
- There is an obligation upon the patient to seek medical help. Whenever a treatment for an illness is assumed to be medically beneficial, there is an obligation upon a patient to undergo such treatment. He who refrains from

doing so is described in Scripture: "And surely your blood of your lives will I require."[17]

- There is an obligation of respect and dignity toward one's fellow man.
- There is a call for solidarity and mutually shared values and duties in society rather than individualism and extreme autonomy.

JEWISH MEDICAL ETHICS VS SECULAR MEDICAL ETHICS

Jewish medical ethics, in terms of the application of Halakhic (Jewish legal) and Jewish ethical principles to the solution of problems, differs from secular medical ethics on four planes:

- the range of discussions and attitudes
- the methods of analysis and discussion
- the final conclusions
- the basic principles[18]

THE RANGE OF DISCUSSIONS AND ATTITUDES

Halakhah addresses all the medical ethical questions that secular medical ethics raises, whether old or new. It also addresses specific medical issues that affect only Jews who observe the precepts of the Torah. The basic Jewish approach is the same for questions related to the terminally ill, abortion, and organ transplantation as for questions related to the treatment of patients on the Sabbath, the laws of seclusion, or the laws of a menstruant woman.

METHODS OF ANALYSIS AND DISCUSSION

Jewish medical ethics analyzes medical ethical questions with the same methods and Halakhic principles used for any Halakhic analysis. It uses basic principles and sources enunciated in the Talmud, Codes of Jewish law, and the responsa literature (collections of rabbinical replies to questions on Jewish law) of all generations. The scientific or medical data are presented, and the relevant Halakhic sources are then applied to the data. It is not always easy to arrive at a Halakhic conclusion regarding a medical question. A far-reaching knowledge of Halakhah as well as an expert and precise understanding of the relevant scientific facts is required in order to arrive at the proper Halakhic conclusion.

Judaism, in general, prefers the casuistic approach to resolve Halakhic questions. This means that one must examine each situation according to the individual circumstances and develop the response according to the specific details and characteristics of that situation, using many of the basic Halakhic rules, regulations, and principles. This is the methodology of the responsa literature and is ideally suited for medical questions where the circumstances differ from patient to patient. By contrast, the current approach of Western secular medical ethics uses a limited number of ethical principles and applies them to all situations involving medical ethical questions.

The Halakhic construct in resolving a medical ethical question is a tripartite one involving the patient and/or family, the physician, and the rabbinic decisor. The patient is obligated to seek the best possible medical care. He has the autonomous right to choose his physician and his rabbinic decisor as well as the right to make his personal wishes known. The physician is obligated to treat the patient and must use the best diagnostic and therapeutic interventions according to his knowledge and judgment. The rabbinic decisor is obligated to understand all the facts of the medical questions, to consider the views presented by the patient and the physician, and then to decide, according to Halakhic principles, how to proceed in any given situation. His decision is binding on both patient and physician. This construct can be termed a religious-paternalistic approach that restricts the patient's as well as the physician's autonomy and that requires acceptance of the Halakhic decision; it negates personal paternalism. This construct, however, applies only to medical situations that have Halakhic ramifications. Pure medical decisions are made by the physician.

Final Conclusions

Halakhah attempts to give final and operative decisions to questions posed to the rabbinic decisor. Judaism is not just an academic discipline. The goal of studying and teaching Jewish medical ethics, as in all other areas of Torah learning, is to put Torah law and ethics into practice. This approach is in contrast to secular medical ethics, which views its function as defining the relevant ethical dilemma, sharpening the focus of the various views, and not necessarily arriving at final and practical conclusions.

Since time immemorial, however, rabbis have differed in their opinions, and the final decision is not always unanimous. This situation is no different than any other normative legal matter. Mechanisms exist in Halakhah to decide among the various opinions. In this respect, there is no difference between a medical question and any other question in any area of Judaic practice or belief.

Basic Principles

There are marked differences in basic principles of Jewish medical ethics compared to those of secular medical ethics. Jewish ethics, including Jewish medical ethics, is based on duties, obligations, commandments, and reciprocal responsibility. The word "right," in its modern sense meaning "I am entitled to it," does not exist in Biblical or Talmudic literature. By contrast, secular medical ethics is based heavily on the concept of rights and autonomy. This minimalistic view justifies human decisions that cannot be criticized as long as they do no harm to others. Judaism, however, requires self-fulfillment based on obligatory and binding moral requirements that are beyond the personal, temporal feeling of individuals; rather, they are founded on values mutually beneficial to society.

Judaism recognizes absolutism only with respect to the Divine source of authority of Jewish law, the supreme authority of the prophets who speak the words of God, and the eternity of Torah. Judaism does not, in general, subscribe to a set of principles and values as absolute imperative categories. Rather, it favors a middle of the road approach, the "path of the golden mean," which is a proper balance between different values or laws in any specific case.[19] The ethical imperative for the average person is to conduct oneself properly with the appropriate balance between opposing values and to avoid extreme positions. Hence, for Judaism there is no definitive value that is absolute so that it takes precedence in every case or situation. Various values have different moral weight, and there is a system for ascribing priorities in specific situations where conflicting values exist. This view is based on the principle that "the Torah was not given to ministering angels"[20] but to ordinary human beings who, by definition, are not perfect. Man's obligation is "to be strong as a lion to arise in the morning to the service of one's Creator."[21] Namely, one must seek and strive to reach Divine understanding and perfection in Torah.

The physician-patient relationship in Judaism is not a voluntary contractual arrangement; it is a Divine commandment and obligation. The patient is commanded to seek healing from the physician and to prevent illness, if possible. The physician is obligated to heal and is considered to be the messenger of God in the care of patients. The patient is not free to decide autonomously to refuse treatment that might be beneficial or save his life. He is prohibited from relying on miracles, and he must do whatever is necessary to heal himself according to standard medical practice.

In Judaism, the value of human life is supreme; therefore, to save a life, nearly all Biblical laws are waived. This approach is in contrast to the secular ethical view, which considers human life to be one of many values and often gives greater weight to "the quality of life." Nonetheless, even in Judaism, the value of human life is not absolute and, in certain rare and well-defined circumstances, other values may supersede it. This, however, does not in any way diminish the supreme value of human life in Judaism. The emphasis placed on human life in Judaism exceeds that of most other religions.

The four basic principles widely accepted in secular medical ethics today are also accepted as important values in Judaism, but they do not receive the same weight in the Jewish tradition.

The principle of autonomy, which is dominant in Western secular medical ethics, is modified in Judaism. Judaism asserts that man was created in the image of God[22] and that all people are, therefore, considered special and equal.[23] Thus, Judaism requires that people must respect and help one another. Judaism also accepts a degree of patient autonomy in the physician-patient relationship. However, in certain situations in which autonomy conflicts with other fundamental principles of Judaism, such as the obligations to preserve one's health and life, to avoid harming others, and to do good for others, the Halakhah may be in direct conflict with autonomy.

In Judaism, man is said to have free will and choice. This does not mean that he is permitted to choose to live immorally or to violate Torah laws. A person is commanded to live within Halakhah, and, thus, his autonomy and free choice are restricted. Decision making in areas that do not involve Halakhah can be totally autonomous. However, in every life situation in which there is a clear Halakhic position, any Jew, be he the physician or the patient, must always act within the parameters of Halakhah and not on one's own inclinations and desires.

The principles of beneficence and non-maleficence are clearly defined axioms in Judaism that prohibit the intentional harming of another person either physically, emotionally, or financially or by defamation or by an attack on objects owned by others. In addition, Jewish law clearly requires not only the avoidance of harm to others, but the active doing of good to others. Sometimes, punishment is inflicted for not doing so. This approach is in contradistinction to secular law and ethics that usually only require one to avoid harm to others but do not obligate one to do good for others. While acts of kindness are considered praiseworthy, they are not specifically required in secular law and ethics as they are in Jewish law. Thus, coming to the aid of a stranger ("good Samaritanism"), considered a supererogatory act in most Western societies, is obligatory in Judaism.

Risky Treatments

Early Sources

The main source dealing with the question of high-risk medical situations is the Biblical story of the four lepers who sat at the city gates during the war between Israel and Aram.

> And they said one to another, "Why sit here until we die? If we say: we will enter the city when the famine is in the city we shall die there; but if we remain here, we die also; therefore, let us fall into the camp of Aram; if they permit us to live, we shall live, and if they kill us, we shall die.[24]

The Talmud concludes from this episode that one may forfeit short time survival (*Chayei Shaah*) if there is any hope for long life (*Chayei Olam*).[25]

Another Talmudic source seems to contradict this rule. It is stated, "One should desecrate the Sabbath by removing debris from a collapsed house in order to save a life of the hour."[26] This denotes the concept that even a very short span of life takes precedence over one of the strictest laws in Judaism, namely desecration of the Sabbath.

The answer to this contradiction is given by commentaries of the Talmud. In both instances, we do whatever is good for the patient with a "life of the hour." In the case of desecrating the Sabbath, if one does not interfere, the patient will certainly die. In the case of treating a terminally ill patient, if one does not take the chance of treatment, he surely will die.[27]

Jewish Ethical Rulings

The Jewish principle concerning a decision to use a dangerous treatment is as follows: A patient who is estimated to die within 12 months (life of the hour) because of a fatal illness is permitted to undergo a treatment that, on the one hand, may extend his life beyond 12 months but, on the other hand, may hasten his death (shorter than the natural course of his lethal illness).[28]

There are, however, several limitations to this ruling. Some rabbis limit this permissive ruling to situations where the chances of success with the proposed treatment are at least 50 percent.[29] Other rabbis rule that even if the chances for success are less than 50 percent, it is permissible to undergo the treatment provided the success rate is at least 30 percent.[30] Yet, other rabbis rule that as long as there are any chances for prolonging life, it is permissible because it is being done for the patient's benefit with the chance, even remote, of prolonging the patient's life.[31]

Some rabbis limit this permissive ruling to situations where the intent of the treatment is curative. If the treatment, however, will not eliminate the illness or the danger but will merely postpone the danger and death, it is prohibited if the treatment itself may actually hasten the patient's death.[32]

One is not *obligated* to undergo a dangerous treatment, but one is *permitted* to do so. However, if the chances of success are very high, one is obligated to submit to potentially life-saving treatment.[33]

The permissibility to endanger oneself to achieve a cure from an illness applies if the treatment or surgery is absolutely necessary and without which the patient will die. However, if there is doubt, if the patient might survive without the treatment, and if the treatment itself might hasten death, it is prohibited to endanger oneself.[34]

Some rabbis write that if a patient is not in danger of dying from an illness but is suffering terribly, and if a treatment can relieve the suffering but may cause the death of the patient, it is prohibited to use it.[35] Other rabbis rule that in such a situation, one should not instruct the patient to undergo that dangerous treatment, but if the patient requests it, it is permissible in order to alleviate his suffering.[36]

The permissibility to forfeit a short life expectancy to achieve more prolonged life applies only with the patient's consent. That consent is valid and is not considered a form of attempted suicide. Neither is a refusal to submit to treatment considered an act of suicide; the patient has the right to refuse a dangerous procedure.[37]

Some rabbis rule that in all situations where a significant risk exists, one may proceed with the treatment if both a majority of expert consultants agree and the rabbinical advisor approves.[38] In all situations where a permissive ruling is granted for a patient to endanger his short life expectancy, the ruling should be arrived at after careful reflection and with the approval of the rabbinic authorities acting on the recommendation of the most expert physicians.[39]

This difficult decision-making process occurs quite frequently in modern medicine, as so vividly exemplified by the personal case of Dr. Michael E. DeBakey described by Dr. George Noon in this book. A similar situation occurred a few years ago in Israel. Rabbi Yosef Shalom Eliashiv, currently the most authoritative Halakhic decisor, suffered from a dissecting aneurysm of the aorta. Given his medical condition at the time, the Israeli physicians' risk assessment was much higher than 30 percent. Since Rabbi Eliashiv's Halakhic position prohibits risky treatment beyond 30 percent,[40] he refused the operation. Only when an expert team from abroad offered a stent procedure with a risk of 30 percent did Rabbi Eliashiv agree to undergo this operation, and the results were excellent.

NOTES

1. Psalms 34:15.
2. Leviticus 19:18.
3. Jerushalmi Nedarim 9:4; See further Maimonides' *Deot* 6:3.

4. Shabbat 31a.
5. Leviticus 18:21.
6. Yoma 86a; Maimonides' *Yesodei Hatorah* 5:1ff.
7. Deuteronomy 6:18.
8. Isaiah 56:1.
9. Amos 5:15.
10. Micah 6:8.
11. Leviticus 19:16.
12. Deuteronomy 16:20.
13. Genesis 1:28.
14. Proverbs 2:20.
15. Proverbs 3:17.
16. See the portrayal of righteous and proper in Isaiah 33:15; Ezekiel 18:5–9; Psalms chapters 1, 15, and 24; Job, chapters 29–31.
17. Genesis 9:5
18. See Avraham Steinberg, "A Jewish Perspective on the Four Principles" in *Principles of Health Care Ethics: Vol. 1 – Theological Developments is Bioethics*, ed. Raanan Gillon (Chichester: Wiley Publishing, 1994), 65–73; and Avraham Steinberg, "Jewish Medical Ethics" in *Bioethics Yearbook*, vol. 1, ed. B.A. Brody et al. (New York: Springer, 1991), 179–199.
19. As stated by Solomon (Ecclesiastes 7:16–18) and by Maimonides (*Deot* 1:3–5); see also Chapter 4 of Maimonides' *Eight Chapters* and *Lechem Mishneh, Deot* 1:4.
20. Berachot 25b.
21. *Shulchan Aruch, Orach Chayim* 1:1.
22. Genesis 9:6.
23. See Avot 3:17; Sanhedrin 38b; Malachi 2:10.
24. 2 Kings 7:3–4; Sanhedrin 107b.
25. Abodah Zarah 27b; *Shulchan Aruch, Yoreh Deah* 155:1.
26. Yoma 85a.
27. *Tosafot* Avodah Zara 27b.
28. *Shulchan Aruch, Yoreh Deah* 155:1; Responsa *Shevut Yaakov*, Part 3 #75; Responsa *Mishpat Kohen* #144:3; Responsa *Achiezer, Yoreh Deah*, #16:6; Responsa *Iggrot Moshe, Yoreh Deah*, Part 3 #36; Responsa *Tzitz Eliezer*, Part 4 #13 and Part 10 #25:17.
29. *Mishnat Chachamim* on *Hilchot Abodah Zarah* in *Yavin Shemuah*, end of #39; Responsa *Tzitz Eliezer*, part 10 #25, Chapt. 5:5.
30. Personal communication from Rabbi Y. Zilberstein, in the name of Rabbi Y.S. Eliashiv.
31. Responsa *Achiezer, Yoreh Deah* #16:6; Responsa *Iggrot Moshe, Yoreh Deah*, Part 2 #58 and Part 3 #36.
32. Responsa *Iggrot Moshe*, Part 3 #36.
33. Ibid.
34. Responsa *Binyan Av*, part 1 #50:1.
35. Responsa *Iggrot Moshe*, Part 3 #36.
36. *Mor Uketzia* #328; Responsa *Tzitz Eliezer*, part 4 #13:2.
37. Responsa *Iggrot Moshe*, Part 3 #36; Responsa *Tzitz Eliezer*, Part 4 #13 and #13:2 and Part 10 #25:17.
38. Responsa *Shevut Yaakov*, Part 3 #75.
39. *Mishnat Chachamim*, on *Hilchot Abodah Zarah*, end of #39, cited in Responsa *Achiezer*, #16:6.
40. See note 30.

AFTERWORD

I was very lucky my parents could afford to send me to Germany for a surgical residency during the Third Reich. I was looking forward to a career in surgery and research in a medical institution, and the training available in the United States was mostly pretty mediocre. Although there were some spots of good surgery like the Mayo Clinic and the University of Pennsylvania in the 1930s, there was simply nothing in the United States comparable to the prestigious universities in Germany.

Dr. Alton Ochsner, under whose influence I had come as a medical student at Tulane, had recommended that I train with Dr. Martin Kirschner in Heidelberg. He had done the first successful pulmonary embolectomy and was very precise, very disciplined, and a great teacher of surgery. Dr. Kirschner also had a good animal research laboratory; therefore, I got some research training as well.

My German hosts treated me very nicely. I had taken German in high school but did not speak it fluently, so I took a course, and, by the time I left, I was doing pretty well. I had a small circle of friends, and we would occasionally go out and have a beer, but I was pretty busy with my work. I developed valuable associations, and it was an extremely rewarding period in my life.

I was not aware of Nazi influence during my training in Germany. I was not attuned to political developments. I knew that there was a general policy that had developed against the Jews. On one occasion, my closest friend and colleague, who was German, told me how sorry he was that a friend of his left the country but that his friend was probably happier elsewhere. I also heard that one of my mentors had departed for Switzerland because he was Jewish, but I just wasn't interested enough at the time to learn more about this. I thought there might have been prejudice against Jewish doctors, but it never occurred to me at the time that this was official policy.

After finishing my surgical training in Germany, I returned to Tulane, where I worked full time in the Department of Surgery. When the war began, I applied to the service as soon as I could and was assigned to a new surgical consultant division in the surgeon general's office under the command of Colonel Rankin, a physician acquaintance of Dr. Ochsner, and spent five years in his office. While I was in the surgeon general's office, I heard indirectly that Germany had concentration camps. At first I didn't connect them with the Jewish people—that came later when we got into Germany and the army overran the camps. Word came to the surgeon general's office that the army had found a prison filled with Jewish youngsters, women, and older men and that they were being prepared for their death sentences. We were absolutely horrified! Germany is not a savage country. It is a civilized country. What are they doing? I couldn't believe it. I just couldn't believe it. We got the facts and it was true, and the awareness of the physician involvement made it all the more horrendous. I still have difficulty believing it. I lived with these people. They were not savages; they were nice folks. What happened to them? That's what horrified me—my God, if this could happen to them, it could happen to us. During wartime, there is a modification of the culture and habits of a country. Even though you might be a very peaceful person, in wartime you develop a militant character. But at no time does that militant character drive you to savagery. The attitude of the highest German authorities against the Jews was pure savagery.

I don't think it will ever happen again. The reason I say this is because it is so horrendous. You have to have the combination of a crazy person leading a lot of passive people and that combination is rare. Hitler could have stopped when he got to Paris, but he chose to go to war with the United States, which was the biggest mistake he made. For example, while I was moving forward with Patton's army, we had to bring in gas for tanks by air. Engineers quickly built airports, and every minute DC3's would come down, land, and give up their gas. The Germans couldn't possibly beat us because we could produce more materiel in one or two days than they could produce in a month. You would think that a man at the head of a great country like Germany would have advisors to tell him that it's suicide to go against the United States.

In the 63 years since the end of the war, America has replaced Germany as the world's leader in medicine. This transformation came about primarily because of the outstanding medical research and development in prestigious American universities, the National Institutes of Health, and the private sector that began in the 1940s while I was in the surgeon general's office.

I had testified before Senator Lister Hill's committee. (His father was a surgeon, who was so impressed with Lister that he named his son Lister.) Senator Hill developed an interest in medical research. He took a liking to me, so I encouraged him to expand America's medical research capabilities. He put money into research, including the expansion of the National Institutes of Health, which was the beginning of the high American standards in biomedical research.

I am most familiar with research in cardiovascular surgery. I never accomplished all the things I wanted to accomplish in research, such as finding the cause of arteriosclerosis. On the other hand, I did a number of firsts, operated on more than 60,000 patients, and helped to usher in the development of cardiovascular surgery as a useful specialty. I look back with some pleasure and pride on the treatment of my thoracic aneurysm, a fatal cardiovascular disease that I survived only because of a successful operation by George Noon—an operation that I had first described many years before.

America must continue to pursue excellence in medical research and clinical investigation. A university's faculty must be research minded because without research you are not going to get new knowledge. You may not get it with research, but you can be absolutely certain that without research you get nothing.

One of the greatest recent developments in medicine has been the Human Genome Project, a purely American project. When the genome is developed to the point where any general practitioner can identify the genetic origin of a patient's disease in his office, there will be a tremendous gain in knowledge and a great benefit to that patient. While it is possible that the genome will be used as a discriminatory tool and that we will make the same mistakes that the eugenicists of the twentieth century made, I have no fear that we will because Americans are religious and they are reasonable. I would rather have a friend who is religious than one who is not because I have a greater trust in a person who believes because I think it is more reasonable to believe than not to believe.

If I were Jewish, I would be a little concerned about the savagery that occurred in Germany, but I would take solace in the fact that we are probably past that stage. Germany had a combination of circumstances that will probably never happen again. I don't think it is possible anymore for a whole society to suddenly become savage, to suddenly go crazy. I would hate to think that this age-old prejudice is not being finally put to bed. In medicine, I know it has.

Michael E. DeBakey, M.D.
May 16, 2008

Appendix A: Additional Information

Most of the 23 contributors to this book were lecturers in the Michael DeBakey Medical Ethics Lecture Series sponsored by the Holocaust Museum Houston during the 2007–2008 exhibit How Healing Becomes Killing: Eugenics, Euthanasia, Extermination. Videos of all 33 lecturers, including the superb speakers not included in this book, can be viewed online at http://www.utexas.edu/cola/centers/scjs/med-ethics/lectures.php, a site made available through the Schusterman Center for Jewish Studies in the College of Liberal Arts at the University of Texas in Austin.

The exhibit catalog, *How Healing Becomes Killing: Eugenics, Euthanasia, Extermination*, can be purchased through the Holocaust Museum Houston website, https://www7.ssldomain.com/hmh/store_item.asp?id=30.

INDEX

U

V

W

Y

Z

CPSIA information can be obtained at www.ICGtesting.com
Printed in the USA
LVOW10s0634071213

364295LV00009B/26/P

9 780230 621923